I0787956

Microfluidic MEMS/NEMS, Sensors and Devices

Editors:

P. Vanýsek
Northern Illinois University
DeKalb, Illinois, USA

D. Cliffel
Vanderbilt University
Nashville, Tennessee, USA

P. Hesketh
Georgia Institute of Technology
Atlanta, Georgia, USA

A. Khosla
Concordia University
Montreal, Québec, Canada

Sponsoring Divisions:

 **Sensor**

 Physical and Analytical Electrochemistry

 New Technology Subcommittee

Published by
The Electrochemical Society
65 South Main Street, Building D
Pennington, NJ 08534-2839, USA
tel   609 737 1902
fax   609 737 2743
www.electrochem.org

ecstransactions ™

Vol. 58, No. 40

Copyright 2014 by The Electrochemical Society.
All rights reserved.

This book has been registered with Copyright Clearance Center.
For further information, please contact the Copyright Clearance Center,
Salem, Massachusetts.

Published by:

The Electrochemical Society
65 South Main Street
Pennington, New Jersey 08534-2839, USA

Telephone 609.737.1902
Fax 609.737.2743
e-mail: ecs@electrochem.org
Web: www.electrochem.org

ISSN 1938-6737 (online)
ISSN 1938-5862 (print)
ISSN 2151-2051 (cd-rom)

ISBN 978-1-62332-172-7 (Softcover)
ISBN 978-1-60768-528-9 (PDF)

Printed in the United States of America.

Preface

The papers included in this issue of *ECS Transactions* were originally presented in the symposium "Microfluidic MEMS/NEMS, Sensors and Devices", held during the 224th meeting of The Electrochemical Society, in San Francisco, CA from October 27 to November 1, 2013.

ECS Transactions, **Volume 58, Issue 40**
Microfluidic MEMS/NEMS, Sensors and Devices

Table of Contents

Preface   *iii*

Real Time Diagnostic Point of Care by Amperometric Immuno-Biosensor Kit by Flow   1
Technology
   H. E. Braustein, I. E. Braustein

Corrosion Sensors: II. Wireless Miniature Corrosion   19
   W. H. Smyrl

(Invited) Self-Assembly of Aligned One Dimensional ZnO Nanorod Arrays   33
on Electron Beam Lithographically Patterned Templates for Sensor Applications
   A. K. Pradhan, C. Samantaray, M. Arslan, H. Dondapati, T. Birdsong, K. Santiago,
   D. Biswal

(Invited) Micromolding of NiFe and Ni Thick Films for 3D Integration of MEMS   41
   M. Cortés, J. Moulin, M. Couty, T. Peng, O. Garel, T. H. N. Dinh, Y. Zhu,
   M. Souadda, M. Woytasik, E. Lefeuvre

Sm-Co Thick Films Micromolding   55
   J. Moulin, M. Woytasik, D. Belghiti, K. Chouarbi

Design and Modeling of a Novel Two Dimensional Nano-Scaled Force Sensor Based   65
on Silicon Photonic Crystal
   T. Li, L. Li, W. Song, G. Zhang, Y. Li

Author Index   75

Facts about ECS

The Electrochemical Society (ECS) is an international, nonprofit, scientific, educational organization founded for the advancement of the theory and practice of electrochemistry, electronics, and allied subjects. The Society was founded in Philadelphia in 1902 and incorporated in 1930. There are currently over 7,000 scientists and engineers from more than 70 countries who hold individual membership; the Society is also supported by more than 100 corporations through Corporate Memberships.

The technical activities of the Society are carried on by Divisions. Sections of the Society have been organized in a number of cities and regions. Major international meetings of the Society are held in the spring and fall of each year. At these meetings, the Divisions and Groups hold general sessions and sponsor symposia on specialized subjects.

The Society has an active publication program that includes the following:

Journal of The Electrochemical Society — (JES) is the leader in the field of electrochemical science and technology. This peer-reviewed journal publishes an average of 550 pages of 85 articles each month. Articles are published online as soon as possible after undergoing the peer-review process. The online version is considered the final version and is fully citable with articles assigned specific page numbers within specific issues. The date of online publication is the official publication date of record.

Journal of Solid State Science and Technology — (JSS) is one of the newest peer-reviewed journals from ECS launched in 2012. JSS covers fundamental and applied areas of solid state science and technology including experimental and theoretical aspects of the chemistry and physics of materials and devices. Articles are published online as soon as possible after undergoing the peer-review process. The online version is considered the final version and is fully citable with articles assigned specific page numbers within specific issues. The date of online publication is the official publication date of record.

Electrochemistry Letters — (EEL) is one of the newest journals from ECS launched in 2012. It is dedicated to the rapid dissemination of peer-reviewed and concise research reports in fundamental and applied areas of electrochemical science and technology. Articles are published online as soon as possible after undergoing the peer-review process. The online version is considered the final version and is fully citable with articles assigned specific page numbers within specific issues. The date of online publication is the official publication date of record.

Solid State Letters — (SSL) is one of the newest journals from ECS launched in 2012. It is dedicated to the rapid dissemination of peer-reviewed and concise research reports in fundamental and applied areas of solid state science and technology. Articles are published online as soon as possible after undergoing the peer-review process. The online version is considered the final version and is fully citable with articles assigned specific page numbers within specific issues. The date of online publication is the official publication date of record.

Electrochemical and Solid-State Letters — (ESL) was the first rapid-publication electronic journal dedicated to covering the leading edge of research and development in the field of solid-state and electrochemical science and technology. ESL was a joint publication of ECS and IEEE Electron Devices Society. Volume 1 began July 1998 and contained six issues, thereafter new volumes began with the January issue and contained 12 issues. The final issue of ESL was Volume 16, Number 6, 2012. Preserved as an archive, ESL has since been replaced by SSL and EEL.

Interface— *Interface* is an authoritative yet accessible publication for those in the field of solid-state and electrochemical science and technology. Published quarterly, this four-color magazine contains technical articles about the latest developments in the field, and presents news and information about and for members of ECS.

ECS Meeting Abstracts— *ECS Meeting Abstracts* contain extended abstracts of the technical papers presented at the ECS biannual meetings and ECS-sponsored meetings. This publication offers a first look into the current research in the field. ECS Meeting Abstracts are freely available to all visitors to the ECS Digital Library.

ECS Transactions— (ECST) is the online database containing full-text content of proceedings from ECS meetings and ECS-sponsored meetings. ECST is a high-quality venue for authors and an excellent resource for researchers. The papers appearing in ECST are reviewed to ensure that submissions meet generally-accepted scientific standards. Each meeting is represented by a volume and each symposium by an issue.

Monograph Volumes — The Society sponsors the publication of hardbound monograph volumes, which provide authoritative accounts of specific topics in electrochemistry, solid-state science, and related disciplines.

For more information on these and other Society activities, visit the ECS website:

www.electrochem.org

Real Time Diagnostic Point of Care by Amperometric Immuno-Biosensor Kit by Flow Technology

Harold E. Braustein[a]*, Isabella E. Braustein[b],

The " George Wise" Life Science Institute, Molecular Microbiology and Biotechnology Department, Tel Aviv University, Tel Aviv, Israel[a]
Jerusalem District Health Office, Ministry of Health, Jerusalem, Israel[b]
*E-mail: Harold.Braustein@Gmail.com

There is a growing need for virus-detecting sensors with improved sensitivity and dynamic range, for applications including disease diagnosis, pharmaceutical research, agriculture and homeland security. We report a novel electrochemical biosensing method for improving the sensitivity for detection of the bacteriophage virus MS2, using nanoporous oxirane-derivatized beads. These beads are a commercial polymethyl-metacrylate (PMMA) polymer that has extremely high surface area to volume ratio, making it an ideal platform for surface based sensors. We have developed and evaluated a method for covalent bioconjugation of antibodies and biological support to polymeric beads. The resulting Solid State Kits (SSK) were used to selectively capture enzyme-labeled MS2 viruses from different solutions, enabling detection of a viral concentration of as low as 10 plaque-forming units per milliliter (pfu ml^{-1}) by measuring the current (A) from the exposed SSK beads to the enzymatic reaction electrons movement not clear. The kit is connected to a "home made" designed micro- flow system, that exhibits sensitivity and dynamic range similar to the ELISA immuno- liquid array-based assay while outperforming protein micro-array methods.

Immuno-Amperometric techniques, using nano- Bio-Polymers Solid Phase Disposable Kit, were used to measure and thus to validate the accuracy of novel technology for virus concentration determination. These work demonstrate the utility of immuno-electrochemical techniques for use in environmental-health quality assurance measurements of viruses.

Introduction

Chemical and biological sensors are analytical devices that can provide quantitative or semiquantitative information (1-36) by exploiting a chemical or biological recognition element (37-40) In the sensors, the recognition element is either integrated within or is closely associated with a transducer interface that can convert chemical, physical, or biological interactions into a measurable output. During the past two decades, sensors, especially biosensors, have become necessary for detecting different analytes such as explosives (41,42), proteins (43,44) DNA (45,46), cancer markers (47,48), bacteria (49,50), viruses (51,52) and toxins (53,54) in food processing, environmental monitoring, clinical diagnostics, and the fight against bioterrorism (3,19,20). As a result, there is a pressing need for the development of target-selective sensors. Viruses, which can specifically infect their host, are being employed to overcome such a challenge. Although the natural function of the viruses is to store and transport genetic material, they have recently been demonstrated to act as templates for the synthesis and assembly of nanomaterials (23-27), as vehicles for targeted drug and gene delivery (28,29), and as probes for sensing and imaging (30-32). Three types of viruses may be used as platforms for these applications: bacteriophages, plant viruses, and animal viruses (33,34). The use of viruses in sensing is being actively developed, and many opportunities are left for scientists to explore in this area. For example, MS2 phage is a stimulant for biothreat viruses, such as smallpox, and thus developing a method for detecting this phage is of paramount important. An electrochemical immunoassay based on the application of paramagnetic beads was designed to detect MS2 phage by amperometry (132). A biotinylated rabbit anti-MS2 IgG was first attached to a streptavidin-coated bead, which then captured the phage. A rabbit anti-MS2 IgG-β-gal conjugate was then attached to the other side of the phage. The β-gal initiates the conversion of PAPG into p-aminophenol (PAP), which can be oxidized electrochemically to p-quinone imine (PQI). The current arising from this reaction is directly proportional to the concentration of the antigen in the sample, and can be measured by rotating disk electrode (RDE) amperometry by using an interdigitated array electrode. The detection limit of this electrochemical method can reach 3.2×10^{10} viral particles per ml (RDE amperometry at an applied potential of + 290 mV versus Ag/AgCl and 3000 rpm) (53). A fully automated fluidic system can be designed for a bead-based immunoassay with amperometric detection of the MS2 phage and ovalbumin (OVA) (54).

Eupergit® [a copolymer of N,N '-methylenebis-(methacrylamide), glycidyl methacrylate, allyl glycidyl ether, and methacrylamide] deserves special attention because it exhibits good chemical, mechanical, and other properties such as low swelling tendency in common solvents, ability to handle high flow rates in column, an excellent performance in stirred batch reactors, *etc.* (30).

The presence of epoxy groups in Eupergit® provides a binding site for enzymes (32, 33, 34). Since the sugar residues in glycoenzymes are often not required for their activities, a novel more efficient strategy, which includes binding to polymers via activated carbohydrate moieties, seems to be very promising (33,34).
This type of binding results in favorable orientation of the immobilized enzyme, since its active site region moves to further distance from the support, facilitating interaction with a substrate.

Thus, exist different approaches for covalent immobilization of enzyme on Eupergit® C are available (commonly used).

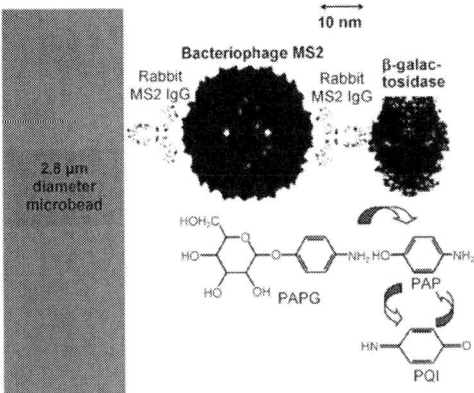

Fig.1 Schematic illustration of the bead-based immunoassay for bacteriophage MS2. Biotinylated rabbit anti-MS2 IgG (biotin-1° Ab) is attached to the streptavidin-coated microbead. Bacteriophage MS2 is sandwiched between the biotin-1° Ab and rabbit anti-MS2 IgG labeled with β-galactosidase (conjugate). The enzyme label converts *p*-amino-phenyl-β-D-galactopyranoside (PAPG) into *p*-aminophenol (PAP). PAP is oxidized through a two-electron reaction to *p*-quinone imine (PQI) at an electrode with an applied potential of + 290 mV versus Ag/AgCl and PQI can be reduced to PAP at −300 mV (132).

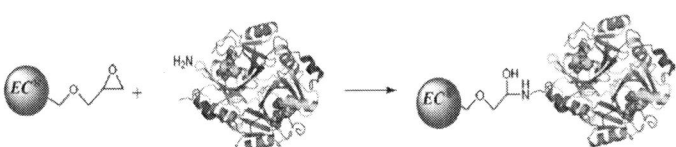

Fig.2 : Schematic linkage reaction between the oxirane group of the Eupergit polymeric bead and a enzyme by –NH2 (amino) group.

The purpose of the work described here was to combine SWCNT nano-amplification techniques using SPE (Screen Printed Electrodes) with an oxirane-polymeric bead-based immunoassay, packed on a disposable kit (the plastic micro-tip Fig. 2) to detect the MS2 bacteriophage. The disposable kit is connected in a micro-flow system, projecting a model for viruses quantification and recognition, in different medium studied.

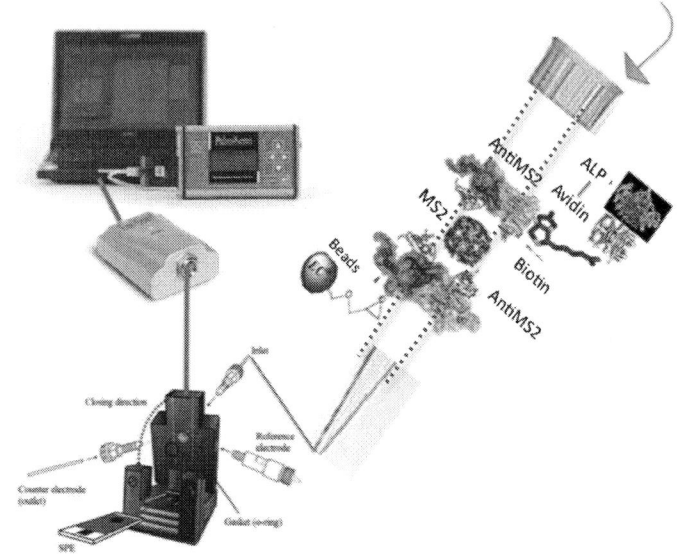

Fig. 3 The scheme of micro flow system with the Solid State kit for the immuno-assay for quantification of the MS2 bacteriophage as a model for viruses analysis in liquid phase

2. EXPERIMENTAL

2.1. Materials

αMS2, from MS2 bacteriophage originating from E Coli, was prepared by Water Quality Research Laboratory, National Public Health Laboratories, Ministry of Health, Tel-Aviv at Tetracore Inc. Rehovot. MS2 at concentration of 10^{10} cfu ml^{-1} was obtained from same laboratory.

Eupergit C 250 L, alkaline phosphatase, Biotin, Avidin-ALP, 1-naphtyl phosphate disodium salt, potassium chloride, and other reagents were purchased from Sigma Aldrich, Israel, with high purity. All the agars were bought from Seat Carmel Israel.

2.2. Biomolecules

Anti-E. Coli O157:H7 monoclonal antibodies (Abs) were purchased from Remel (Lenexa, KS, USA). Anti-MS2 polyclonal Abs was produced at Tetracore Inc. Rehovot and precipitated from the serum by ammonium sulphate (Pierce Protocol).

Half the quantity was linked to Avidin-labelled alkaline phosphatase (ALP).

2.3. Apparatus

The electrochemical quantifications of the MS2 bacteriopahage at different concentrations were made with a homemade apparatus. The enzymatic product, resulted as eluent from the solid state micro-column-tip, was checked on the WE of SPE MWCNT modified.

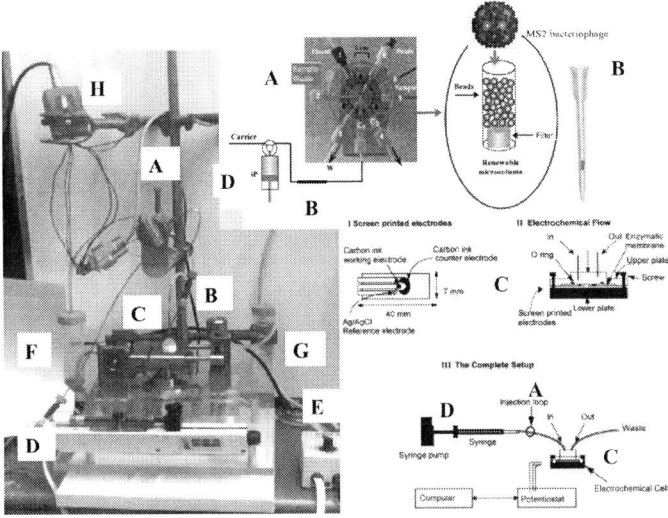

Fig.4-Left: the Microflow system and its components: A- injector; B-the microtip column; C-the SPE "house" containing the EC; D-syringe of the syringe pump; E- speed flow regulation of the syringe pump; F- EC Waste; G- Injector(A) waste; H- potentiostat (EmStat – on the picture); Right up- Injector and microcolumn packed with beads Scheme; Right -I. the Screen printed electrode; II-the EC; III-the microflow system-complete setup amperometrical checked on the "home made" electrochemically "house" for the SPE. The electrochemical "house" consist from a press system, build from stainless steel. The electrode is inserted in the designated place, in the apparatus, sealed by a 3 mm diameter Teflon stopper that has in/out capillary tubing entrance/exit for the liquid phase. The working buffer is continuously circulated in the system, by a 2.5 ml syringe pump, with electro- mechanic speed regulation (0-100 µl min^{-1}). The Teflon stopper sealed the WE of the SPE, by an O-ring (gasket), providing a 30-µl electrochemical cell. The whole system is connected to a 6 valves HPLC injector for analytes /reagents analysis injection (Fig. 3 A).

Equal aliquots from the enzymatic reaction were analyzed using the Micro-Flow Electrochemical System built in our laboratory, connected to a mini potentiostat (PalmStat or EmStat) bought from PalmSense Inc., Netherland. The instrument is controlled and the data is recorded and displayed by the computer system, [Compaq(HP)

Laptop, for laboratory work or a (HP) Palm Handled Device], equipped with electrochemical PalmSense Inc.software, for field work.

3. Assay of MS2 Activity

3.1. Bacterial strains and bacteriophages

Heat killed E. Coli O157:H7, purchased from KPL Inc. (Gaithersburg, MD, USA) at a concentration of 10^{10} CFU ml^{-1}, were used in this study. The heat treatment did not affect the recognition of the strain by the anti-E. Coli O157:H7 antibodies (Abs) as confirmed with ELISA tests.

E. Coli K12, a laboratory strain, was used to evaluate the biosensor selectivity. A stock culture was prepared by growing the cells overnight in Luria Bertani (LB) broth at 37 °C with continuous agitation, at the end of which the cells were harvested by centrifugation and washed three times in phosphate buffer 0.1 M (PB), pH=6.7. Growth was quantified by plating the bacteria stock culture at ten-fold serial dilutions on LB agar, and counting the number of colonies after overnight incubation at 37 °C.

A bacteriophage T7 (ATCC BAA-1025-B2) stock was prepared using E. Coli (ATCC BAA-1025) as bacteriophage host. Phage concentration was determined by the double layer agar method (Adams, 1959). A bacteriophage stock of 10^9 PFU ml^{-1} was 10-fold serially diluted in PBS for assessment with the fabricated biosensors. MS2 phage was used as negative control in the selectivity studies. A stock of this phage was prepared using E. Coli HS(pFamp)R as host strain for phage replication and the concentration was determined similarly as described above for the T7 phage.

3.2. Preparation of bacterial lysates

The bacterial cells were sonicated for 5 min with pulses of 15 s ON and 30 s OFF using a Virtis sonicator (Virsonic 600). Lysis of the cells was confirmed microscopically. The cell lysates were used immediately.

3.3. Microbial Stock Production and Microbiological Detection Methods:

The recovery and downstream detection of several microorganisms were used in the laboratory studies evaluating the tangential flow ultrafiltration system.

3.3.1 MS2 Bacteriophage – Cell culture validation of the dilutions used in electrochemical, optical assay and double agar overlay method for MS2 bacteriophage enumeration:

1.   MS2 stock generation: To generate high titer (10^{13} pfu) bacteriophage stocks, a bacterial host (*i.e.* Escherichia Coli Famp) was incubated in a flask with tryptic soy broth (TSB) plus 1% ampicillin/streptomycin solution at 37°C with shaking for 4 hours to produce a log phase growth. The bacterial host at log phase growth was then used to propagate MS2 using a soft agar overlay method. 7.5 ml of prepared bacterial host and

10 ml of diluted MS2 stock were added to a sterile tube containing 0.5 ml of 0.7% tryptic soy agar (TSA). The contents of the tube were gently mixed and then carefully poured onto four Petri dishes containing 1.5 ml of 1.5% TSA+1% ampicillin/streptomycin solution. The plates are then incubated at 37°C, for 16-18 hours. Using a cell scraper, MS2 was harvested from the plates by gently scraping the top, soft agar layers into a 50cc tube. Next, PBS is added to a total volume of 2.3 ml chloroform (2.3 ml) was then added. After mixing by vortex for 5 minutes at maximum speed the suspension was centrifuged at 4,000xg for 30 minutes in order to separate the MS2 from the extracellular materials and soft agar.

The MS2-containing aqueous phase was carefully removed so as not to disturb the interphase. To reduce the formation of aggregates, the stock was filtered through sequentially smaller pore low protein-binding filters (0.45, 0.22 and 0.1 micrometer), pretreated with 0.5 ml of 0.1% Tween 80 and 0.5 ml PBS. Aliquots of MS2 bacteriophage stocks were stored at –80°C.

MS2 bacteriophage were enumerated using the double agar overlay procedure in U.S. EPA Method 1602. 7.5ml of prepared bacterial host and 10ml of MS2 stock serially diluted in a buffer (PBS, 0.01% Tween 80 and 0.001% Antifoam-A) were added to a sterile tube containing 0.5 ml of 0.7% tryptic soy agar (TSA). The contents of the tube are mixed gently by 'rolling' between the analyst's palms and then carefully poured onto a Petri dish containing 1.5 ml of 1.5% TSA+1%ampicillin/streptomycin solution. The sample is spread evenly over the surface of the plate by gently and quickly swirling the plate. The plate, which solidifies within 30 seconds, is then inverted and incubated at 37°C for 16-18 hours. During the incubation time, the host bacteria form a confluent lawn over the surface of the Petri plate allowing the phage particles that are present in the sample to attach to and infiltrate the bacterial host cells. The MS2 bacteriophage then replicates within the host cells eventually causing it to lyse the bacterium. The destruction of the bacterial cells that make up the confluent lawn result in clear areas known as plaque forming units. The concentration of bacteriophage present in the sample is determined by visually counting the plaques (Figure 4).

The results were checked in parallel with the results obtained in the electrochemical measurements, to correlate the dilutions used in the two methods considered.

3.4. Electrochemical measurements quantification

In a typical reaction mixture, 100 mg work portion of Eupergit Beads were incubated directly with αMS2, at a concentration range of 0.5 – 15 μg / 1 mg beads, at room temperature (RT), for 1 – 8 hours. The beads were washed by centrifugation at 12,000 rpm with 0.1 M pH=7 PB containing 0.15 % Tween 20. After a blocking step by 16 h incubation with 1.5 % - 3 % Bovine Serum Albumin (BSA) or skim milk, following a PBS wash, the samples of MS2 in their respective media (buffer, blood serum, wastewater or urine) at known concentrations were added. αMS2-Biotin was labeled with ALP-Avidin (alkaline phosphatase). The beads were weighted in approximately 40 columns (on 10 μl filter plastic tips, of 2.5 mg each) and used for the MS2 quantifications. To each column prepared were injected 40-60 μl of 1 mM 1-naphtyl phosphate, freshly prepared before the experimental work). The resulted eluent (product of the enzymatic reaction) was checked at 300 mV, using 0.1 M pH=8.5 – 9.2 PB with 0.1 M KCl, on the WE (Working Electrode), previous nano- modified by 3x 3 µl multi walled carbon nano-tubes-MWCNT (40 g l^{-1}-in water suspended), previous prepared by 3 times incubations layers at 37°C, for ½ hour. The enzymatic product was amperometrical checked on the micro-flow system.

3.5. Monoclonal phage enzyme-linked immunosorbent assay (ELISA)

3.5.1. Optical biosensor measurements quantification

3.5.1.1. Microbiological quantifications Dilutions

An optical immuno-assay ELISA was designed using the same materials and samples, by checking the optical densities at 450 nm. Bacteria were grown overnight at 37°C and 2 µl samples were transferred into 200 µl of medium in fresh blocks. Once the OD600 reached 0.4, 25 µl aliquots of KM13 helper phage, each containing approximately 10^9 virus particles, were added to each well to initiate super infection. After a further 1 h growth, the plates were centrifuged at 1800 rpm for 10 min and the supernatant aspirated from each well. Individual cell pellets were resuspended in 200 µl of 2YT Broth containing ampicillin and kanamycin and the plates were incubated at 30°C with shaking overnight to allow viral replication.

The plates were then centrifuged at 10,000 rpm for 10 min and the supernatants were transferred directly to an ELISA plate pre-coated with the antigen(αMS2). For coating of the ELISA plates, MS2 coat protein or neutravidin (2 mg/ml in PBS; 100 µl/well) was added and plates were incubated overnight at 4 °C.

The wells were washed 3 times with 0.1M PBS (250 µl/well), containing 0.05% Tween-20 and blocked for 1 h with 1.5% skim milk in PBS (MPBS). The wells were again washed 3 times with PBS, and phage-containing supernatants were then added for 1 h at room temperature. For the thermal stability experiments, phage-containing supernatants were heated to 70°C for 30 min and quenched on ice prior to incubation with the antigen-coated wells.

Binding of the phage was detected using a polyclonal anti-MS2 Ab, conjugated to alkaline phosphatase (ALP; ABCam, USA).

3.5.2. Direct MS2 Optical biosensor by ELISA

Standard 96 wells ELISA plates are incubated with αMS2 polyclonal, 50 µl each well. The plate was incubated ON (over night) with 1.5% BSA or skim milk, 300 µl/well. To the wells were added 100 µl calibrators and samples at different concentrations (10 – 10^{10} pfu ml^{-1}), followed by addition of αMS2-PAb-ALP labeled.

3.6. Determination of Optimum Temperature and pH

Optimum temperatures were determined by changing temperature in the range of 10 to 65 °C while keeping the substrate concentration constant (0.1 mM). Also, pH optimization was carried out by changing the pH range between 7 and 9 at constant temperature (27 °C).

3.7. Operational Stabilities and Shelf Life.

The operational stability of the enzyme biosensor based on the Solid State kit was determined at optimum activity conditions using Micro Tip Kits in 30 activity immuno-assays per day, on the mini-flow system designed, using the same SPE nano-modified, for all the assays. The shelf life of the micro-columns tip kits were investigated by performing activity measurements within 25 days.

4. Results and discussion

Binding of the proteins to Eupergit C may be achieved by reaction of their amino, thio-, or hydroxy- groups with oxirane moieties on the surface of the matrix to obtain highly active preparations (2,3,11). However, matrix immobilization may cause partial or complete inactivation. In most cases, this effect may result from modification of amino acid residues essential for activity, improper orientation of the immobilized molecules, induced conformational changes, or restrictions induced by multipoint attachment of the protein molecules to the surface of the matrix (11).

In the following study the Escherichia Coli/MS2 system was chosen for a number of reasons. MS2 is harmless to humans and yet possesses many of the same features as its eukaryotic-infecting viral cousins, and as a result may aid in our understanding of RNA viruses in general. It can be cultured quickly, cheaply, and safely, making it easy to work with. Furthermore, the genome-scale metabolic model of E. Coli is the most exhaustive one to date. These factors combine to make the E. Coli/MS2 model system ideal for such a study.

We tested the sensitivity and dynamic range of the sensor by exposing it to a series of solutions with viral concentrations ranging from 10^{10} to 10 plaque-forming units per ml (pfu ml^{-1}). As shown in Fig. 4, we obtain MS2 detection in cell culture produced from serial dilutions of MS2 using the DAL method (Adams, 1959).

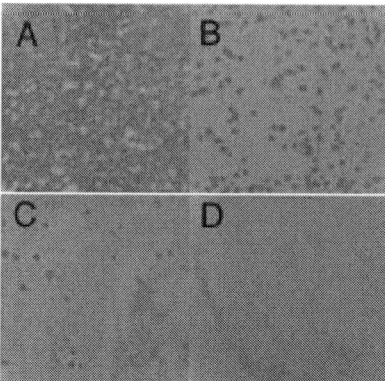

Figure 5. Typical cell culture results of dilution series of MS2 using the DAL plaque assay method 5.A. – cell culture of stock solution (approx. 10^{10} pfu ml^{-1}); 5.B. – cell culture of dilution of stock solution (approx. 10^5 pfu mL^{-1}); 5.C. – cell culture of dilution of stock solution (approx. 10 pfu mL^{-1}); 5.D. – cell culture of dilution of stock solution (approx. 0 pfu ml^{-1})- the blank solution

4.1. Immunoassay Results

In the present study we propose an electrochemical bead-based sandwich enzymatic immunoassay with alkaline phosphatase as the enzyme label which converts 1-Naphtyl Phosphate substrate to electrochemically-detectable 1-naphtol ion (see Fig. 5).

At the end of the second incubation with enzyme substrate in the assay procedure (carried out by injection through the microtip column), a 30 μl eluent drop was transferred to the micro-flow system injector. As illustrated in Figure 5, the drop containing the enzymatic product (and unreacted substrate) was placed by the flowing buffer directly on the array (WE), and close to the fine wires used for reference and auxiliary electrodes carefully inserted, in the 30 μl EC. There is no limitation of this arrangement because the micro-drop tends to pass the SPE, straight to WE (Working Electrode) rapidly. Therefore, measurements were made at a medium speed (125 μl min^{-1}) immediately after the drop was injected. The electrode response to a range of concentrations of MS2 virus as a function of time is shown in Figure 6. After an initial decline, the currents stabilized after approximately 40 s. The general current stabilize may be due to buffer continuous circulation in the micro-flow system. With the finger gap being in the nanometer range, collection efficiency should be very high making the anodic current only slightly larger than the cathodic current at steady state. The anodic current is larger than the cathodic current, which was attributed to some oxidation of unreacted 1NP(1-naphtyl phosphate) substrate present in the drop.

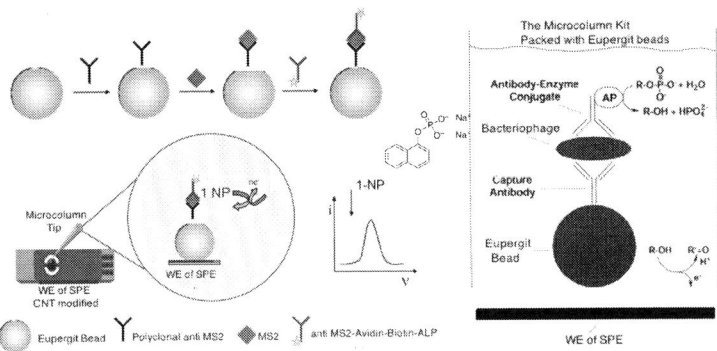

Fig. 6. Left-the immunoassay sandwich built on the Eupergit beads, the electron transfer from the enzymatic reaction of Alkaline phosphatase and 1-naphtyl phosphate disodium salt; Right-The enzymatic reaction of alkaline phosphatase labeling on αMS2, on the immuno-assay linked to the Eupergit beads, in the microcolumn kit.

The MS2 samples diluted in different media (wastewater, serum blood, urine) were analyzed in triplicates at 300 mV, due the presence of ALP as enzyme used in the assay. Although calibration could be done by measuring the current anywhere, in field or in the laboratory, from about 1 to 3 min, for 6 samples (5 calibrators and one sample). The calibration curve for an MS2 immunoassay shows increasing current with concentrations between 10 and 50 pfu ml^{-1}. The general shape is characteristic of a sandwich immunoassay, well-defined because of our ability to obtain more points for more concentrations and repeat each measurement (at least three repetitions). The results indicate that lower detection limits may be possible. One limitation of an enzyme label system that converts substrate to a detectable product is that no concentration of substrate is optimal for all analyte concentration ranges not clear. At high analyte concentrations, a higher substrate concentration is required to ensure that product formation is not reduced due to substrate depletion.

This is especially important in microliter volumes. When detecting very low concentrations, the background signal caused by the non-enzyme-catalyzed conversion of substrate to product, as well as the slight electrochemical activity of the substrate, limits the detection sensitivity.

Optimization of substrate concentration and sample volume would lower the limit of detection further. Therefore, a curve with at least 5 points and triple measurements for each point was obtained, illustrating the possibility of making sensitive nano-or even pico-molar concentrations measurements that cannot be done with traditional techniques. The information in Table at Fig. 7B illustrates the sensitivity that can be gained by redox cycling in essentially the same immunoassay procedure for MS2 using different quantities of αMS2/αMS2-ALP linked to the beads.

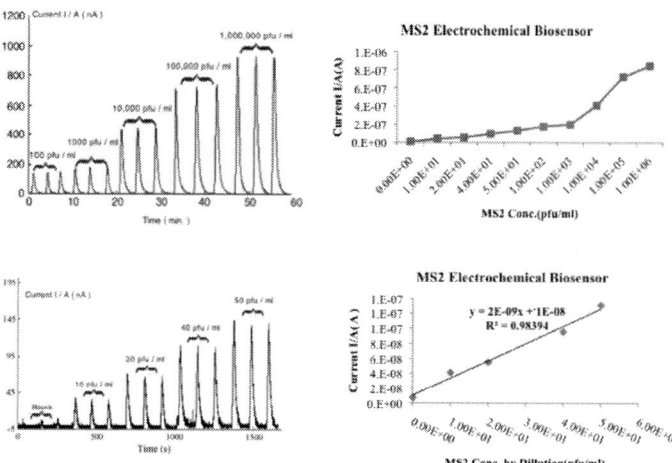

Fig.7 Results of MS2 bacteriophage immuno-assay, using the Solid State microtip column, in the mini flow system. Up – Quantification of 10^2-10^6 pfu ml^{-1} MS2 bacteriophage. Down - Quantification of 0 -50 pfu/ml MS2 bacteriophage A-D

In addition, a smaller detection volume is required for measurement with the SPE nano-modified, incorporated on the mini flow system, further reducing the detectable amount of analyte. In this work, several optimizations were done, to establish the optimal work conditions. Several linked concentrations at different values of the αMS2 and αMS2-ALP were done to receive amplified signals and also, we tried to optimize the amounts of biological support, to receive a economical assay kit, as can be seen in Fig. 7 (Table-Right up).

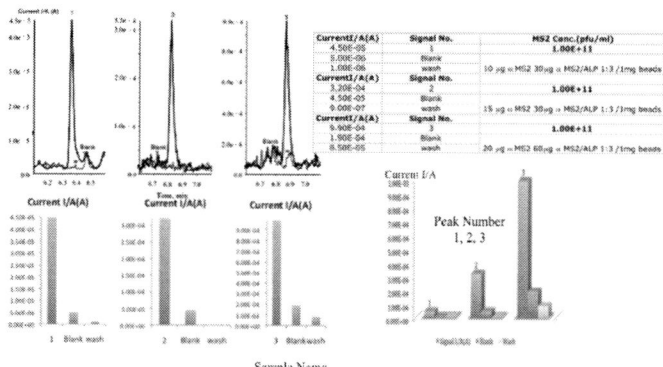

Fig.8 Optimization of work conditions for MS2 10^{11} pfu/ml quantification. Peak 1 result of immuno-assay using 10 μg αMS2 30 μg a MS2/ALP 1:3 /1mg beads; Peak 2 result of immuno-assay using 15 μg α MS2 30 μg α MS2/ALP 1:3 /1mg beads; Peak 3 result of immuno-assay using 20 μg α MS2 60 μg αMS2/ALP 1:3 /1mg beads Fig7 A-D.

The immunoassay is amplified as a direct result of the optimal concentration of biological support linked to the beads. The microtip columns are packed with small quantities of beads (2.5 mg). The results of the quantifications taken as accurate are those that are approximately 3 times higher than the blank. After each injection of the bacteriophage concentrations used (calibrator or sample), the system is washed by injecting double distilled water, to get rid of most salts from the buffers and other chemicals accumulated on the capillary tubes of the mini-flow system.

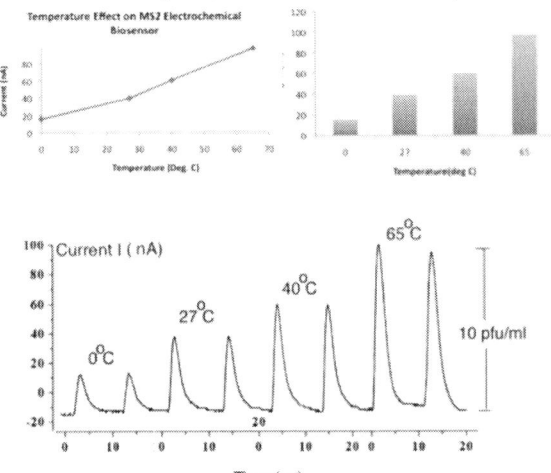

Fig. 9 Optimization of working conditions for MS2 10 pfu/ml quantification. Results of duplicates of the immuno-assay at different work temperatures.

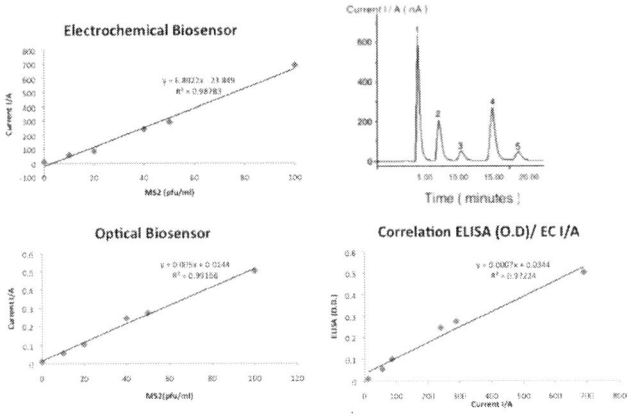

Fig.9 Comparison between MS2 quantifications results obtained with the Electrochemical immunoassay in the mini flow system and results of ELISA, using the Solid State micro-tip column kit.

The immuno-electrochemical results of different MS2 bacteriophage concentrations were checked on an optical biosensor for MS2 (described in methods) by ELISA. Assay results on blood serum confirmed good accuracy of the mini fluidic array by showing excellent correlations with individual standard ELISA assays (Fig. 9). These results also demonstrate the high selectivity of the one-protein immunoarray protocol, which was able to provide accurate results at clinically relevant concentrations in the presence of hundreds of additional proteins at much higher concentrations in human serum. The high sensitivity of the assay extending to 10 pfu ml^{-1} levels may help facilitate selectivity by allowing high dilution of the serum, although even in the heavily diluted serum there will be hundreds of proteins at higher concentrations than the analytes. In all calibrations, a single Micro tip column was used for one standard concentration of each analysis, then a fresh kit was inserted into the device for assessing additional standards and samples. Good kit-to-kit reproducibility is illustrated by the small error bars (Fig. 6). Detection limits (DL) were measured as 3 times the average SD above the zero bacteriophage control.

Conclusions

A highly sensitive immunoassay for MS2 bacteriophage was obtained using a micro tip Solid State kit interdigitated array nanoelectrode. The increase in sensitivity can be attributed to the improved amplification of the redox cycling compared to the hydrodynamic flow generated by the polymeric kit, and the faster cycling obtained using the nano-modified screen-printed electrode. Detecting biological organisms in realistic sample matrices such as wastewater, serum blood or urine, is significantly more complicated than detecting redox-active analytes in buffer. Although the present micrometer scaled kits are packed with very small quantities of beads, (number/sample) their high sensitivity in this preliminary work offers promise for miniaturized electrochemical sensors.

To summarize, we have developed highly sensitive and selective polymeric solid-state kit, SWCNTs-based chemiresistive biosensors for detection of bacteriophage MS2 using a facile and affordable technology that can be employed in most laboratories. A limit of detection of 10 PFU/ml was achieved for the bacteriophage MS2. Through its lower limit of detection, the advantages of this biosensor rely on its simple measurement performance, allowing measurements to be performed in a real-time manner at room temperature. The integration of nanostructures such as SWCNTs into microelectronic devices in combination with the use of specific recognition molecules such as PAbs has shown promising results for the next generation of truly miniaturized, sensitive, and cost-effective biosensors. Incorporation of microfluidic technology is expected to lower the limit of detection of these devices and increase reproducibility. Since the sensing area of our device is of few micrometers in size, recirculation of the sample may increase the chance for a single particle to reach this area.

A real-time bead-based electrochemical immunoassay has been developed to detect bacteriophage MS2, a biothreat agent stimulant. The detection methods described in this work compare well with other MS2 immunoassays that have been recently reported. Rowe *et al.* (35) have demonstrated an array biosensor that uses a sandwich immunoassay format and fluorescent labels to detect MS2 at 400 ng ml^{-1}. McBride *et al.* (36) reported a detection limit of 3 ng ml^{-1} MS2, using a bead-based assay and a flow cytometer for detection. Although the assay detected with the electrochemical microflow system using a solid state kit is more sensitive, it is more likely that the fluidic channel with SPE electrode, SWCNT nano-modified detection, can be developed into a hand-held device to be used in the field. The consistent next step would be to include automated bead preparation within the micro-fluidic system device.

Our model method falls nearer to the lethal range of hemorrhagic fever viruses (for example), developing sensitivity between 10 and 10^{10} viruses (pfu ml^{-1}). Although improvements in detection limit are needed for an early warning system, the reported method can selectively detect a virus of interest, assuming the appropriate antibody is available.

Acknowledgements

We thank Dr. Klementiy Levkov for building the micro-flow system and precious assistance during the project.

References

1. Espy M. J., Uhl J.R., Sloan L. M., Buckwalter S. P., Jones M. F., Vetter E. A., Yao J. D., Wengenack N. L., Rosenblatt J. E., Cockerill F. R. 3rd, Smith T. F. Real-time PCR in clinical microbiology: applications for routine laboratory testing. *Clin. Microbiol. Rev.* 19(1): 165-256 (2006).
2. Aljaro, C.G. Bangar, M. A. Baldrich, E. Munoz, F. J., Mulchandani, A., Conducting polymer nanowire-based chemiresistive biosensor for the detection of bacterial spores, *Biosens. Bioelectron.*, vol. 25, p.2309-2312(2010)
3. García-Aljaro C., Muñoz-Berbel X., Toby A. Jenkins A., Anicet R. Blanch, and Muñoz F. X., Surface Plasmon Resonance Assay for Real-Time Monitoring of Somatic Coliphages in Wastewaters, *Appl. Environ. Microbiol.* 74 (13):4054–4058.doi: 10.1128/AEM.02806-07 PMCID: PMC2446531(2008).
4. Garcia-Aljaro C, Munoz-Berbel X, Munoz F. J.: On-chip impedimetric detection of bacteriophages in dairy samples. *Biosens. Bioelectron.*, 24:1712-1716 (2009).
5. Iijima, S. Helical microtubules of graphitic carbon. *Nature* 354 (6348): 56-58. Bibcode:1991Natur.354.56I. doi:10.1038/354056a0 (1991).
6. Merkoçi A, Pumera M, Llopis X, Pérez B, del Valle M, Alegret S. TrAC, *Trends. Anal. Chem.* 24:826–838. doi: 10.1016/j.trac.2005.03.019 (2005).
7. Trojanowicz, M., Trac-Trends *Anal. Chem.* 25 (5), 480–489 (2006).
8. Allen B. L., Kichambare P. D., Star,A. Carbon Nanotube Field-Effect-Transistor-BasedBiosensors, *Adv.Mater.*19,1439–1451, DOI: 10.1002/adma.200602043 (2007).
9. Lin Y, F Lu, Y Tu, and Z Ren. Glucose Biosensors Based on Carbon Nanotube Nanoelectrode Ensembles. *Nano Letters* 4(2):191-195 (2004).
10. Wang J., Liu G., & Jan M.. Ultrasensitive Electrical Biosensing of Proteins and DNA: Carbon-Nanotube Derived Amplification of the Recognition and Transduction Events. *J. Am. Chem. Soc.*, Vol. 126, No. 10 , pp. 3010 – 3011, ISSN 0002-7863 (2004).
11. Maehashi K., Katsura T., Kerman K., Takamura Y., Matsumoto K., Tamiya E. Label-free protein biosensor based on aptamer-modified carbon nanotube field-effect transistors, *Anal. Chem.*, 79, pp. 782–787(2007)

12. Villamizar R. A., Maroto A., Rius F. X., I. Inza and Figueras M. J., Fast detection of *Salmonella Infantis* with carbon nanotube field effect transistors *Biosens. Bioelectron.* 24 (2), 279-283 (2008).
13. Villamizar R. A. & Maroto A. & Rius F. X., Rapid detection of Aspergillus flavus in rice using biofunctionalized carbon nanotube field effect transistors, *Sens. Actuator B-Chem.* 136 (2), 451–457 (2009).
14. Lim J.H., Phiboolsirichit N., Mubeen S., Rheem Y., Deshusses M. Mulchandani A., Nosang A., Myung V, Electrical and Sensing Properties of Single-Walled Carbon Nanotubes Network: Effect of Alignment and Selective Breakdown, *Electroanalysis* 22 (1), 99–105 (2010).
15. Lakshmi N. Cella,Pablo Sanchez, Wenwan Zhong,Nosang V. Myung,Wilfred Chen, and Ashok Mulchandani, NANO APTASENSOR FOR PROTECTIVE ANTIGEN TOXIN OF ANTHRAX, *Anal. Chem.* 82, 2042–2047 (2010).
16. Crowther, Viruses and the Development of Quantitative Biological Electron Microscopy, IUBMB Life 56, 239–248. (2004)
17. Curry A, Appleton H, Dowsett B., Application of transmission electron microscopy to the clinical study of viral and bacterial infections: present and future, *Micron* 37, 91–106 (2006).
18. Ksiazek T. G., Erdman D., Goldsmith C. S., Zaki S. R., Peret T., Emery S., Tong S., Urbani C., Comer J. A., Lim W., Rollin P. E., Dowell S. F., Ling Ai-Ee, Humphrey C. D., Shieh Wun-Ju, Guarner J., Paddock C. D., Rota P., Fields B., DeRisi J., Yang Jyh-Yuan, Cox N., Hughes J. M., LeDuc J. W., Bellini W. J., Anderson L. J., and the SARS Working Group, A Novel Coronavirus Associated with Severe Acute Respiratory Syndrome, *N. Engl. J. Med.* 348, 1953–1966. (2003)
19. Brittan L. Pasloske, Cindy R. Walkerpeach, R. Dawn Obermoeller, Matthew Winkler, and Dwight B. DuBois, Armored RNA Technology for Production of Ribonuclease-Resistant Viral RNA Controls and Standards, *J. Clin. Microbiol.* 36, 3590–3594 (1998).
20. Smith, H. O., Hutchison, C. A., III, Pfannkoch, C. & Venter, J. C., Generating a synthetic genome by whole genome assembly: φX174 bacteriophage from synthetic oligonucleotides *Proc. Natl. Acad. Sci.* USA100 , 15440-15445(2003)
21. Sambrook, J., Russell, D.W., Molecular Cloning a Laboratory Manual. *Cold Spring Harbor Press*, Cold Spring Harbor, NY, pp. 2.38–32.39, 32.44, 32.48–32.51(2001)
22. Davis J. E. , Sinsheimer R. L. The replication of bacteriophage MS2. 1. Transfer of parental nucleic acid to progeny phage. *J. Mol. Biol.* Mar;6:203–207.(1963)
23. Eisenberg, Physical chemistry with applications to the life sciences, *Benjamin/Cummings*, Menlo Park, CA. (1979)
24. Mazzone H. M., CRC Handbook of Viruses: Mass-Molecular Weight Values and Related Properties. CRC Press (Boca Raton), 240 pages (1998).
25. Guinier A., *Ann. Phys.* 12, 161 (1939) Comptes Rendus 223, 31 (1946).
26. Guiltier A. and Fournet G., Small Angle Scattering of X-rays (Wiley, New York,), p. 151 (1955).
27. Jacrot B., Zaccai G., Determination of molecular weight by neutron scattering, ; *Biopolymers* 20, 2413. (1981)
28. Koch M. H., Vachette P., Svergun D. I., Small-angle scattering: a view on the properties, structures and structural changes of biological macromolecules in solution, *Q. Rev. Biophys.* 36, 147–227. (2003)
29. Zheng Y. Z., Webb R., Greenfield P. F., Reid S.: Improved method for counting virus and virus like particles. *J. Virol. Methods*, 62:153-159 (1996).

30. Katchalski-Katzir, K., and Kraemer D.M.: Eupergit$^{(R)}$ C, a Carrier for immobilization of enzymes of industrial potential. J. *Mol. Catal. B: Enzym.* 10 157-176. (2000)

31. HernaizaM. J. , Crouta D. G. Immobilization/stabilization on Eupergit C of the β-galactosidase from B. circulans and an α-galactosidase from Aspergillus oryzae, *Enzyme Microb. Technol.* 27, 26–32. (2000)

32. Rocchietti S., Urrutia A.S.V., Pregnolato M., Tagliani A., Guisan J.M., Fernandez-Lafuente, Influence of the enzyme derivative preparation and substrate structure on the enantioselectivity penicillin G acylase, *Enzyme Microb. Technol.* 31, 88–93 (2002).

33. Wirz B., Barner R., Huebscher J., Facile chemoenzymic preparation of enantiomerically pure 2-methylglycerol derivatives as versatile trifunctional C4-synthons, *J. Am. Chem. Soc.* 58 3980–3984 (1993).

34. Hilala N., Kochkodanb V., Nigmatullina R., Goncharukb V., Laila Al-Khatiba, *J. Membr. Sci.* 268, 198–207.

35. Rowe C. A., Tender L. M., Feldstein M. J., Golden J. P., Scruggs S. B., MacCraith B. D., Cras J. J., and Ligler F. S., *Anal. Chem.* 71 3846–3852 (1999).

36. McBride M. T., Gammon S., Pitesky M., Multiplexed liquid arrays for simultaneous detection of simulants of biological warfare agents. *Anal Chem* 75:1924–1930. (2003)

37. Turner A. P. F, Karube I., Wilson G.S., In: Biosensors: Fundamentals and Applications. Turner A. P. F., editor. Oxford University Press; Oxford: 1987. p. 770.

38. Thévenot D. R.,Toth K., Durst R. A., Wilson G.S. *Biosens. Bioelectron.*2001;16:121.

39. Turner APF. *Science.* 2000;290:1315.

40. Willner I. *Science.* 2002;298:2407.

41. Steinfeld JI. *Annu. Rev. Phys. Chem.* 1998;49:203.

42. Yinon J. TrAC *Trends Anal. Chem.* 2002;21:292.

43. Diercksa AH, Ozinskya A, Hansen CL, Spotts JM, Rodrigueza DJ, Aderema A. *Anal. Biochem.* 2009;386:30.

45. Tsou PH, Chou CK, Saldana S, Hung MC, Kameoka *J. Nanotech.* 2008;19:445714.

46. Gagnon Z, Senapati S, Gordon J, Chang HC. *Electrophoresis.* 2008;29:4808.

47. Guo Q, Yang X, Wang K, Tan W, Li W, Tang H, Li H.*Nucleic Acids Res.* 2009;37:e20.

48. Brower V. *J. Natl. Cancer Inst.* 2009;101:11.

49. Gokarna A, Jin LH, Hwang JS, Cho YH, Lim YT, Chung BH, Youn SH, Choi DS, Lim JH. *Proteomics.* 2008;8:1809.

50. PandaBR,SinghAK,RameshA,ChattopadhyayA. *Langmuir.* 2008;24:11995.

51. Luo PG, Stutzenberger FJ. *Adv. Appl. Microbiol.* 2008;63:145.

52. MacCuspie RI, Nuraje N, Lee S-Y, Runge A, Matsui H. *J. Am. Chem. Soc.* 2008;130:887.

53. Thomas JH, Kim SK, Hesketh PJ, Halsall HB, Heineman WR. *Anal. Chem.* 2004;76:2700.

54. Kuramitz H, Dziewatkoski M, Barnett B, Halsall HB, Heineman WR. *Anal. Chim. Acta.*2006;561:69.

55. Formatting by

Corrosion Sensors: II. Wireless Miniature Corrosion

William H. Smyrl

University of Minnesota Corrosion Research Center Department of Chemical
Engineering and Materials Science Minneapolis, Minnesota 55455

The research reported here will focus on a miniature battery system
that is inactive as assembled ("off"). It is activated when a
damaging fluid flows into the inter-electrode separator ("on"). In
the "on" state, it powers a wireless sensor that transmits a signal to
the base station monitor to announce that corrosion conditions
have been created at the location of the battery – sensor. Further,
the battery-sensor powered the measurement of the time-to-failure
(TTF) of a corroding specimen that was initiated when the fluid
arrived (on), and ended when the specimen failed (off). The on/off
action is a "large" signal that is insensitive to electrical noise from
the ambient surroundings.

Introduction to wireless corrosion sensors

Metals used in engineering structures in modern societies are thermodynamically
unstable in their service environments. The overall objectives of corrosion sensing are
first to discover the presence of a damaging fluid at a given location, and second to
monitor corrosion. The traditional methods to study corrosion of exposed samples
involve electrical connections to external instrumentation, and they are intrusive. Other
"wired" connections, also intrusive, could be optical fibers used for spectroscopic
detection of species, or for corrosion of metal films for example (Smyrl and Butler [1];
Jin, et al.[2]). Publications that discuss probes and sensors with "wired" connections for
corrosion includes (Chawla, et al.[3]; Agarwala [4]; Bakker [5]; Knutson, et al.[6];
Guillaume, et al. [7]; Wade, et al.[8]; Darwin, et al. [9]; Lindquist, et al.[10]; Alam, et al.
[11]; Chauvin, et al.[12]; and Supino, et al.[13]).

To overcome the problems of traditional approaches to characterize the corrosion of
metals, new wireless sensors should be:

(1) small so that they are not intrusive and do not alter local conditions;

(2) inexpensive so that multiple units can be embedded at distributed locations;

(3) tailored to enable unequivocal predictions of the metal corrosion process in
technological conditions, and;

(4) robust to ensure durability before the onset of corrosion.

Most traditional monitoring techniques characterize the behavior of macroscopic
samples either exposed to simulated conditions or exposed in service conditions.
Broomfield, et al.[14] recently reviewed techniques for studying corrosion degradation:

(1) linear polarization resistance (LPR); (2) critical threshold concentrations of
damaging species (e.g. chloride ions, or pH); (3) resistivity of the damaging medium,
and; (4) half cell potential measurements. Methods (1) to (4) average over sample area
(i.e. corrosion rate by LPR), or determine the concentrations or permeation rates averaged

over large dimensions. Other approaches have been described for measurements in industrial conditions by Dean [15], and a new textbook is an excellent resource for a fundamental background on corrosion science (McCafferty, [16]).

With recent developments in microelectronics, micro-sized sensors are enabled as a new technology. Specialized sensors based on wireless communication and RFID technology offer an attractive new opportunity to deploy inexpensive units (Accenture, [17]). Whether they are adapted from passive or active devices, the sensors might be (a) embedded within engineering structures, (b) deployed in occluded regions or in remote locations, or (c) attached to external surfaces of vulnerable materials. In each case, the sensors could signal a component of the corrosion process without altering the conditions that cause corrosion.

Recent embedded sensors have been used either to determine damaging conditions for corrosion, or to measure degradation response to damaging conditions. Wireless threshold sensors were studied by Dickerson, et al [18], in which a sensing steel wire component in a wireless passive sensor was interrogated with an inductively coupled magnetic monitor to determine the corrosion damage of steel in concrete test structures. Corrosion of the wire disrupted the electrical connection and caused the sensor signal to fail (i.e. a time – to – failure (TTF) device). Watters, et al, [19] developed a sensor as a component of a wireless "Smart Pebble" to determine the concentration of chloride ion within the concrete of a bridge deck. Andringa, et al, [20], used a sensor in which the conductivity and temperature inside a concrete structure was monitored to signal the arrival of the damaging fluid (e.g. water and chloride ion) from the external surface. Earlier, Andringa, et al. [21] discussed low-cost sensors that were inductively-coupled with magnetic monitors to determine the onset of corrosion of a steel wire component within the passive (TTF) sensor. Simonen, et al. [22] introduced prototype passive sensors to monitor corrosion in concrete structures. Materer, et al. [23] published the first patent of a passive wireless device for monitoring corrosion. Chen, et al. [24] recently described work on a network of distributed sensors that function with magnetic induction signals. Another approach was described by Yang, et al. [25] for a surface wetness sensor based on interdigitated electrode arrays (IDA) and RFID technology. A different IDA sensor design was used for RFID sensing the conductivity around a sensor exposed to moisture (Ong, et al. [26]).

The discussion here will focus on a miniature battery system that is inactive as assembled ("off"). It is activated when a damaging fluid flows into the inter-electrode separator ("on"). In the "on" state, it powers a wireless sensor that transmits a signal to the base station monitor to announce that corrosion conditions have been created at the location of the battery – sensor. Further, it powered the measurement of the time-tofailure (TTF) of a corroding specimen that was initiated when the fluid arrived (on), and ended when the specimen failed (off). The on/off action is a "large" signal that has low sensitivity to electrical noise from the ambient surroundings.

Experimental approach

Miniature battery assembly The fabrication of miniature Mg-Ag/AgCl battery is described first for the inactive (or "off") state. A schematic of the cell components of a typical miniature cell is shown in Figure 1. This layout has been found to be effective in our previous studies of flexible fuel cells (Wheldon, et al.[27], Lim, et al.[28; 29]). In the present study, a paper substrate separates the two electrodes and is the host for the electrolyte (damaging fluid) when it arrives. The damaging fluid is absorbed into a tab of the separator outside the cell and the fluid then flows by capillary action into the inter-electrode region of the assembly. A DAKO pen (DAKO Corporation, Carpinteria, CA) was used to create hydrophobic channels and regions in the paper separator. The paper used here is either chromatography paper of medium porosity (Whatman #3001-861), filter paper of high porosity (Whatman #1), or treated samples of either. These papers have few components that could contaminate the fluid and compromise the results. The anode electrode studied here was formed from magnesium foil (99.9% purity, Atomergic Chemicals, Farmingdale, NY). The counter electrode was a Cl- ion reversible electrode, Ag/AgCl, that was coated from a fluid ink (AGCL 675 Silver/Silver Chloride, Conductive Compounds, Hudson, NH). The composition of the AGCL ink had approximately equal mass of Ag and of AgCl, from which the coulombic loading of electrodes was calculated.

When the damaging fluid arrived at the electrode region, the cell was activated ("on" state). We will present results for magnesium – silver/silver chloride electrochemical cells that are activated by arrival of potassium chloride solution as the damaging fluid. Other systems under study will be described in later publications.

Testing of water-activated cells

The Ag/AgCl electrodes were coated with the AGCL ink through a template onto the separator and dried for about 3 days at ambient laboratory conditions. The mass of the electrode material was determined by weighing the paper substrate before and after the active material was coated. The mass was measured with an analytical balance to an accuracy of 0.1 mg. Three day drying times were sufficient to achieve a constant final weight. A metal tab from the electrode was formed by coating with a narrow strip of silver ink (PELCO Colloidal Silver Liquid #16031, Ted Pella, Redland, CA) that extended to the edge of substrate. This tab provided a contact and current collector for the electrode. The fabricated electrodes were stored in the dry state in a darkened location. The open circuit potential (OCP) of the coated electrodes was measured vs. a commercial Saturated Calomel Electrode (SCE) with a Solartron 1287 Electrochemical Interface.

Magnesium electrodes were cut from a metal foil sheet. The foils were etched in 0.1N HCl for several minutes to remove accumulated oxides and residual impurities that remained from the commercial forming and rolling processes. The foil was cut to the final size with a tab that extended outside the edge of the cell for connection to the current collector lead.

The miniature cells were assembled from the components as shown in Fig. 1, and treated with a pouch edge-sealer around all sides of the electrolyte region to confine the damaging fluid to the electrode region when it was introduced into the cell from the absorbent paper tab.

Once the (damaging fluid) electrolyte was added to the tab, it flowed rapidly into the cell by capillary action, and the cell was activated ("on" state) within a few seconds. The activation was very stable and reproducible after a few minutes. The open circuit potential of the activated cell was 1.65V.

Impedance analyses were carried out on symmetrical Ag/AgCl – Ag/AgCl cells with the electrolyte soaked (satd KCl) paper separator (thickness of 0.017 cm). A Solartron 1260 Impedance/Gain Phase Analyzer was used with a LabView controller program developed in our laboratory.. The cells were very stable and reproducible.

Constant current (galvanostatic) discharge was carried out on activated Mg – Ag/AgCl cells ("on" state) at several current levels. A Solartron 1287 Electrochemical Interface was used with a LabView software controller developed in our laboratory

(Steinbach [30]). The change from "off" to "on" was triggered by the arrival of the damaging fluid, and this provided the power for a wireless transmitter and monitor system. The battery-sensor was then used to study corrosion measurements in the form of an "on/off" technique to determine the time-to-failure (TTF) of a corroding sample in the damaging fluid. Results and Discussion Testing of the Ag/AgCl Electrode

The open circuit potential (OCP) of the electrode vs a Saturated Calomel Electrode was determined with a Solartron 1287 Electrochemical Interface. The potential was determined in a saturated potassium chloride solution in deionized water (> 17.9 Mohm resistivity). Both electrodes were immersed in the electrolyte in one test, or the SCE was contacted with the wetted paper substrate on which the Ag/AgCl electrode had been coated in other tests. Both measurements gave identical OCPs. The OCP was independent of mass loading. The Ag/AgCl electrodes were very stable and reproducible, and the potential was -0.042V vs SCE in a saturated KCl electrolytic solution (damaging fluid in the present study). The literature value of the OCP for a Ag/AgCl electrode is 0.045 V vs. SCE when tested in saturated KCl solutions at ambient laboratory temperatures (25oC) (Bard and Faulkner [31]). The Ag/AgCl electrodes were very stable and reproducible, and demonstrate behavior that is identical to that of standard Ag/Ag/Cl (Ives, et al. [32]). Impedance measurements of the Ag/AgCl Electrode Impedance measurements of symmetrical electrode cells were carried out on a Solartron 1260 Impedance Gain Phase/Analyzer with Labview software. The electrodes (equal area of 0.36 cm^2) faced one another across a paper separator (thickness of 0.017 cm) wetted with saturated potassium chloride solution. The cells were assembled as shown in Figure 1, except that the Mg electrode was replaced by a second Ag/AgCl electrode. The impedance behavior was very stable and reproducible over time and for several assembled cells. Results are shown in Figure 2. The high frequency impedance shows a limiting resistance, R_{lim}, of 7.09 ohms. This value includes the resistance of the electrolyte filled paper separator, and contact and current collector resistances. The dimensions of the paper separator and electrode area, along with the conductivity of the KCl solution (0.35825 ohm^{-1}cm^{-1}, Nickels and Allmand [33]) yielded the calculated resistance of the electrolyte phase alone to be 0.133 ohms.

The depressed semicircle(s) of the Cole-Cole plot agree with the results of Rhodes and Buck [34], and will not be analyzed further here. The reduction of AgCl to Ag changes the relative amounts of the two, but does not change the electrode potential so long as both are present. This provided the basis to expect the Ag/AgCl electrode to perform well on discharge in the Mg-Ag/AgCl cells.

Discharge characteristics of Mg-Ag/AgCl cells

The Mg-Ag/AgCl cell has the following reactions for cell discharge

$$Mg(s) = Mg^{+2}(soln) + 2\ e^- \qquad \text{(oxidation)}$$
$$AgCl(s) + e^- = Ag(s) + Cl^-\ (soln) \qquad \text{reduction)}$$

The characteristics of the activated ("on") miniature cells have been studied with galvanostatic discharge. The Ag/AgCl electrode limits the discharge capacity in the results given here. The mass loading of the Ag/AgCl material is approximately 50/50 silver/silver chloride solids (MSDS sheets for AGCL liquids, Conductive Compounds). For a typical mass loading of 0.0350 to 0.0920 grams for electrodes with an area of 1.69 cm^2, the loading of AgCl is calculated to be 7.28 x 10^{-5} to 1.19 x 10^{-4} moles/cm^2. The coulombic capacity of the Ag/AgCl electrode ranged from 7.02 to 18.46 coulombs/cm^2.

The loadings were compared to the discharge capacities measured in the cells for a selected current and time. In Figure 3 is shown the discharge time of 600 seconds and a total coulombic charge of -0.078 coulombs. The voltage of the cell was nearly constant at 1.65 V for the entire discharge, as expected for the short time and low current (only about 1% of the total coulombic capacity of the Ag/AgCl electrode was consumed). Another cell was discharged for a longer time as shown in Figure 4. The latter portion of the discharge (1200-1800 seconds) emphasizes that the average voltage was nearly constant even to the end of the selected discharge time. The current of 1 mA is relatively high for the expected service of the cells in wireless operation, but it was sufficient to demonstrate the TTF corrosion measurement.

We conclude that the Mg-Ag/AgCl cell will support a wireless signal for an hour or more to indicate that the battery is "on" when activated. This has been explored in a proof-of-concept series of tests.

Wireless transmission supported by the Mg-Ag/AgCl cell

To transmit a signal to a remote monitor that the damaging fluid has arrived, we utilized a commercial wireless system that is normally powered by 2 AAA batteries (Springfield Wireless Digital Sensor, Model 91905, Wood-Ridge, NJ, transmission at 433 MHz). The system draws about 10 – 15 microamps current from two batteries in series, and transmits a signal about 100 feet to a remote monitor station. The system is normally used as an outdoor temperature sensor, and we have adapted it for sensing the activation of Mg-Ag/AgCl batteries. We note that the results above have shown that the cells have sufficient voltage (2 cells in series) and current to power the operation of the sensing system, so we have adapted it for sensing corrosion.

Two Mg-Ag/AgCl batteries in series turned on the sensor when a damaging fluid arrived to activate the cells. The cells turn on the sensor and this transmits an "on" signal to the base station. There is sufficient power for continuous operation of an hour or more, depending on the coulombic loading of AgCl on the electrodes. This provides a warning that the critical conditions for corrosion have been detected at the location of the battery-sensor.

Wireless Detection of Corrosion with Mg-Ag/AgCl batteries

We now move to the second stage of the study whereby corrosion is monitored – not just the conditions for corrosion. In order to do this, we use a time-to-failure (TTF) measurement. From the time that the damaging fluid arrives to create corrosion until the test specimen is consumed determines the TTF. The specimen is a thin film of copper (100 nanometers thick) evaporated onto a polyester substrate. A strip of the Cu film was connected in series with the batteries in a similar setup to that used to announce the arrival of the damaging fluid. The second sensor continues to report the presence of damaging fluid, until the Cu film is finally consumed by corrosion in the fluid. This breaks the connection for operation of the battery-powered sensor (i.e., change from on to off), and the TTF of 54 ± 8 minutes was measured for multiple samples in saturated KCl.

The samples were exposed to the damaging fluid (saturated KCl) by placing it onto a wetted strip of chromatography paper. Electrical leads were attached to the Cu film sample and it was placed in series with two Mg-Ag/AgCl batteries, and then to the wireless sensor circuitry. The system continued to operate until the film was corroded through which turned off the power and this turned off the signal transmitted to the base station monitor. The batteries had sufficient capacity for the entire period of the TTF test, and this was repeated for each of the multiple copper film samples.

We note that a TTF measurement has both an induction or initiation period and a propagation period for corrosion. The corrosion of the metal film thickness of metal (here 100 nm of Cu) over the TTF period (initiation plus propagation) did not yield a precise corrosion rate during the propagation period. The corrosion rate of metals often changes with time as well. Nevertheless, the TTF method has value as an indication of corrosion by a damaging fluid. We conclude that these measurements show successful monitoring of the TTF of a corroding sample (metallic Cu film) with our Mg-Ag/AgCl battery-sensor.

References

[1] Smyrl, W.H., and M.A. Butler, "Corrosion Sensors: Early detection of corrosion could save billions of dollars," Interface 2, 35 (Winter 1994).

[2] Jin, W. and W.H. Smyrl, "Reflectivity Monitoring of Anion Adsorption with an Optical Fiber Micromirror", in Critical Factors in Localized Corrosion II, Proc. Electrochem. Soc. 95-15, 333-43(1996).

[3] Chawla, S.K., T. Anguish, and J.H. Payer, "Microsensors for Corrosion Control", Materials Performance, May 1990, 68-74.

[4] Agarwala, V.S., "Sensors for Corrosion Detection and Monitoring in Hidden Areas", in Chemical Sensors, Electrochemical Society, Pennington, NJ, 1992.

[5] Bakker, E., "Electrochemical Sensors", Anal. Chem. 76, 3287(2004).

[6] Knutson, T.L., F. Guillaume, W-J. Lee, M. Alhoshan, W.H. Smyrl "Reactivity of surfaces and imaging with functional NSOM," Electrochimica Acta 48, 32293237 (2003).

[7] Guillaume, F., J. Evju, T.L. Knutson, and W.H. Smyrl, "Exploratory Spectroscopic Imaging at Localized Corrosion Sites", J. Electrochem. Soc. 150(6), B262B265 (2003).

[8] Wade, S.A., K.J. Begbie and A. Trueman, "Measuring Atmospheric Corrosion of Industrial Infrastructure by Using Electrical Resistance Corrosion Sensors" Corrosion Control 2007, paper 118, pp 1-8(2007).

[9] Darwin, D., J. Browning, M. O'Reilly, L. Xing, and J. Ji, "Critical Chloride Corrosion Threshold of Galvanized Reinforcing Bars", ACI Materials Journal April 2009, 176183.

[10] Lindquist, W.D., D. Darwin, J. Browning, G.G. Miller, "Effect of Cracking on Chloride Content in Concrete Bridge Decks", ACI Materials Journal December 2006, 467-483.

[11] Alam, M.N., R.H. Bhuiyan, R. A. Dougal, and M. Ali, "Concrete Moisture Content Measurement Using Interdigitated Near-Field Sensors", IEEE Sensors J., 10(7), 1243-1248 (2010).

[12] Chauvin, M., C. Shield, C. French, and W.H. Smyrl, "Evaluation of Electrochemical Chloride Extraction (ECE) and Fiber Reinforced Polymer (FRP) Wrap Technology", MN/RC – 2000-24, June 2000.

[13] Supino, R., and J.J. Talghader, "Micromachined particles for detecting metal-ion concentration in fluids", J. Microelectromech. Sys. 15(5), 1299(2006).

[14] Broomfield, J.P., K. Davies, and K. Hladky, "The use of permanent corrosion monitoring in new and existing reinforced concrete structures", Cement and Concrete Composites 24, 27-34(2002).

[15] Dean, S.W., "Overview of Corrosion Monitoring in Modern Industrial Plants", ASTM STP: 908, 197-219(1986).

[16] McCafferty, E., Introduction to Corrosion Science, Springer, New York, 2010

[17] Radio Frequency Identification (RFID) White Paper, Accenture (2001).

[18] Dickerson, N. P., M. M. Andringa, J.M. Puryear, S.L. Wood, and D.P. Neikirk, "Wireless Threshold Sensors for Detecting Corrosion in Reinforced Concrete Structures", Smart Structures and Materials 2006: Sensors and Smart Structures Technologies for Civil, Mechanical, and Aerospace System, Proc. of SPIE Vol. 6174, 61741L (2006).

[19] Watters, D.G., P. Jayaweera, A.J. Bahr, D.L. Huestis, N. Priyantha, R. Meline, R. Reis, D. Parks, "Smart Pebble: Wireless sensors for structural health monitoring of bridge decks", Smart structures and Materials 2003: Smart Systems and Nondestructive Evaluation for Civil Infrastructures, Proc. of SPIE Vol 5057, 20(2003).

[20] Andringa, M. M., J. M. Puryear, D.P. Neikirk, and S.L. Wood, "In-situ measurement of conductivity and temperature during concrete curing using passive wireless sensors", Sensors and Smart Structures Technologies for Civil, Mechanical, and Aerospace Systems 2007, Proc. of SPIE Vol. 6529, 65293M (2007).

[21] Andringa, M.M. and D.P. Neikirk, N.P. Dickerson, and S.L. Wood "Unpowered Wireless Corrosion Sensor for Steel Reinforced Concrete", IEEE Sensors 2005, 155158 (2005).

[22] Simonen, J. T., M.M. Andringa, K.M. Grizzle, S.L. Wood, and D.P. Neikirk, "Wireless Sensors for Monitoring Corrosion in Reinforced Concrete Members", Proc. of SPIE 5391, 587-596 (2005).

[23] Materer, N.F., and A.W. Apblett, "Passive Wireless Corrosion Sensor", US Patent Application 20090058427, (2009).

[24] Chen, Y., S. Munukutla, P. Pasupathy, D.P. Neikirk, and S. L. Wood, "Magnetoinductive waveguide as a passive wireless sensor net for structure health monitoring", Proc. SPIE 7647, 1-9 (2010).

[25] Yang, C-H., J-H, Chien, B-Y. Wang, P-H., Chen, and D-S. Lee, "A Flexible Surface Wetness Sensor using a RFID technique", Biomed. Microdevices 10, 1047-1054 (2008).

[26] Ong, J.B., Z. You, J.Mills-Beale, E. L.Tan, B.D. Pereles, and K. G. Ong, "A Wireless, Passive Embedded Sensor for Real-Time Monitoring of Water Content in Civil Engineering Materials", IEEE Sensors J., 8(12), 2053-2058 (2008).

[27] Wheldon, J., Woo-Jin Lee, Dong-Ha Lim, Anders B. Broste, Matthew Bollinger and William H. Smyrl, "High-Performance Flexible Miniature Fuel Cell", Electrochem. Solid-State Lett. 2009, 12(5), B86-B89.

[28] Lim, D-H., Woo-Jin Lee, N. L. Macy, and W. H. Smyrl, "Electrochemical Durability Investigation of PtÕTiO2 Nanotube Catalysts for Polymer Electrolyte Membrane Fuel Cells", Electrochem.Solid-State Lett., 12 (9) B123-B125 (2009).

[29] Lim, D-H., Woo-Jin Lee, J. Wheldon, N. L. Macy, and W. H. Smyrl, "Electrochemical Characterization and Durability of Sputtered Pt Catalysts on TiO2 Nanotube Arrays as a Cathode Material for PEFCs", J. Electrochem. Soc., 157 (6) B862-B867 (2010).

[30] Steinbach, Andrew James, "REVERSIBLE PERFORMANCE STABILITY OF POLYMER ELECTROLYTE MEMBRANE FUEL CELLS", Master of Science Thesis, University of Minnesota, 2008.

[31] Bard, A.J. and L.R. Faulkner, Electrochemical Methods, 2nd ed, John Wiley and Sons, NJ, 2001.

[32] Ives, D.J.G. and G.J. Janz, eds .Reference Electrodes, Academic, New York (1961).

[33] Nickels, L. and A.J. Allmand, "Conductivities and Viscosities at 25C of Solutions of Potassium, Sodium, and Lithium Chlorides, in Water and 0.1M Hydrochloric Acid", J.Phys.Chem. 41, 861-872 (1937).

[34] Rhodes, R.K. and R.P. Buck, Impedance Characterization of Anodized Silver/Silver Chloride Electrodes", Anal. Chim. Acta 113 , 55-66 (1980).

Table 1. Recent Studies of Wireless Sensors for Corrosion

Technique	Objective	Active/ Passive	Reference
Magnetic inductive coupling	TTF* of steel wire	passive	Dickerson, et al ;2006
Smart Pebble (Cl⁻ ref electrode)	Critical Cl⁻ ion concentration	active	Watters et al; 2003
Conductivity (κ) and Temperature inside concrete	Critical Cl⁻ ion and water concentration	passive	Andringa, et al; 2007
Inductively coupled magnetic monitor	TTF* of steel wire	passive	Andringa, et al; 2005
Adapted RFID Tag for corrosion	Corrosion monitor	passive	Materer, et al; 2009 (patent)
Network of magnetic induction sensors	Distributed corrosion sensors	passive	Chen, et al; 2010
Surface wetness sensor, κ, IDA**, RFID	Condensed moisture activated	active	Yang, et al; 2008
κ, IDA**, RFID	Condensed moisture activated	active	Ong, et al; 2008

* TTF – Time to Failure
** IDA – Interdigitated Electrode Array

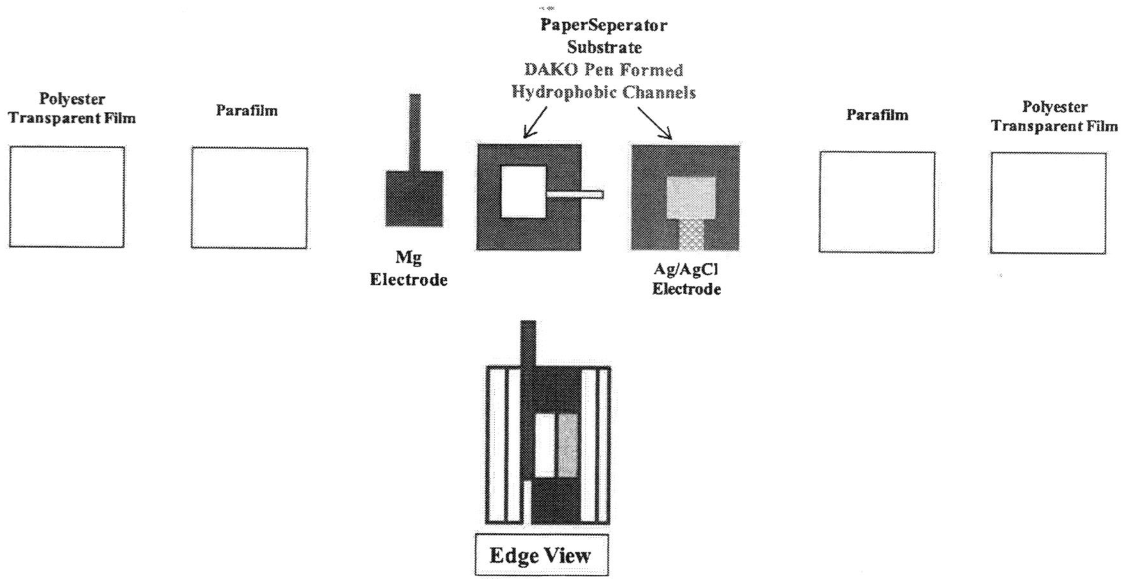

Figure 1. Layout of Miniature Mg-Ag/AgCl Battery Components Assembled Volume of 0.077 cm^3

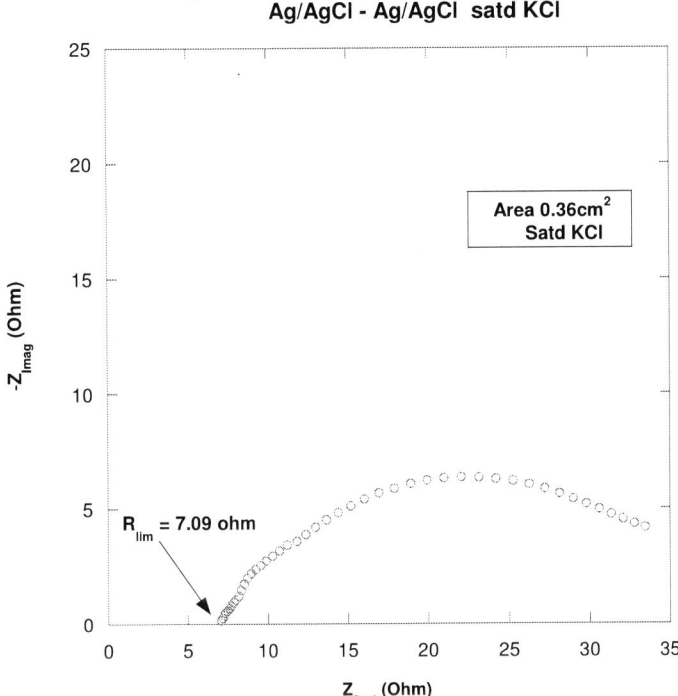

Figure 2. Impedance of Symmetrical Cell
Ag/AgCl - Ag/AgCl satd KCl

Figure 3. Constant Current Discharge Mg-Ag/AgCl Cell Satd KCl

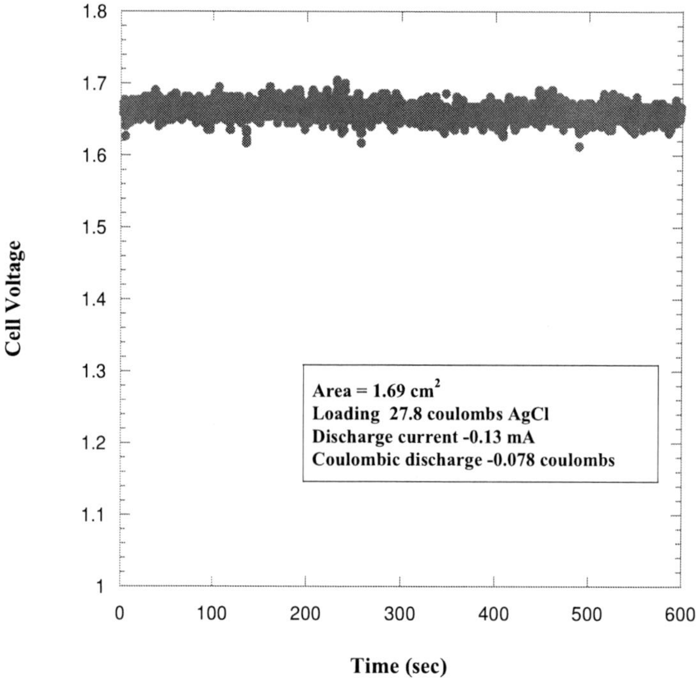
1.8
1.7
1.6
1.5
1.4
1.3
1.2
1.1
1
Cell Voltage
0
100
200
300
400
500
600
Time (sec)
Area = 1.69 cm2
Loading 27.8 coulombs AgCl
Discharge current -0.13 mA
Coulombic discharge -0.078 coulombs

Figure 4. Constant Current Discharge Mg-Ag/AgCl Satd KCl

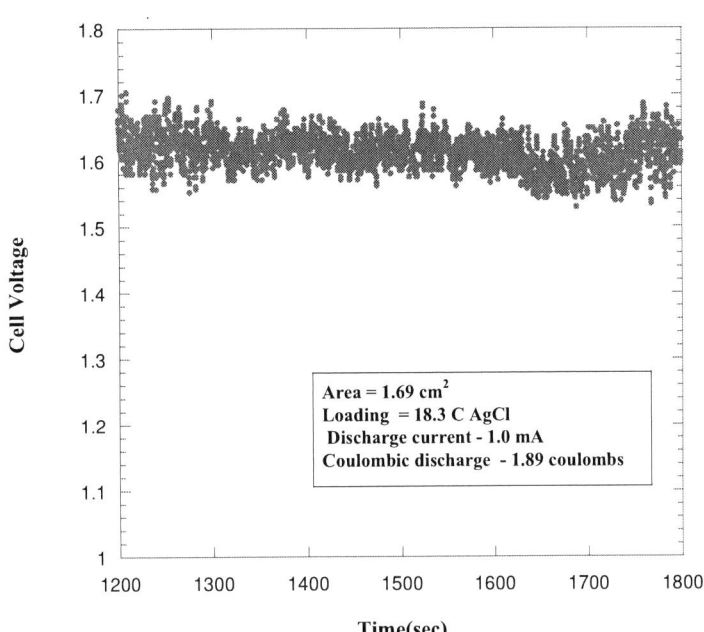

Self-Assembly of Aligned One Dimensional ZnO Nanorod Arrays on Electron Beam Lithographically Patterned Templates for Sensor Applications

A.K. Pradhan*, C. Samantaray, M. Arslan, H. Dondapati, T. Birdsong, K. Santiago, and D. Biswal

Department of Engineering and Center for Materials Research, Norfolk State University, Norfolk, Virginia 23504, USA

Corresponding author: apradhan@nsu.edu

We have demonstrated the attachment of aligned ZnO nanorod arrays on to the electron beam lithographically patterned oxide surface at lower beam energy. At higher beam energy (~ 20 keV), the accumulated negative charges that already built on the surface in due course of irradiation restricts the growth of nanorod arrays. This causes a negative shielding potential close to the surface at micron level, and completely unfavorable for the attachment of the negative ZnO carriers. However, at comparatively lower beam energy (~ 5 keV or less), the secondary electrons are responsible for the pattern with the irradiation zone centered by the local positive field. This allows negatively charged ZnO nanorod arrays to grow at lower beam voltages, which initiate site selective attachments. This protocol of site selection is very useful for various sensor fabrications.

Introduction

A major goal of nanotechnology is to couple the self-assembly of nanostructures and molecular nanostructures with nanofabrication, so-called bottom-up and top-down fabrication methods which can register individual nanostructures to electronically integrate them into functional devices. One approach is to use nanolithography to make templates onto which discrete nanocomponents can self-assemble. Lithographically patterned templates can also be used to create hierarchical order of the nanostructures which can self-organize themselves and their internal dimensional feature can significantly be smaller than those of the original template to serve as scaffolds for the assembly of still smaller components (1). Self-assembly of such nanomaterials in the form of nanoparticles (2, 3), carbon nanotubes (4, 5) and nanowires (6), organic compounds (7), single and double stranded DNA molecules (8-10) or DNA nanostructures (11) were performed on lithographically patterned substrates.

Recently, one-dimensional (1D) array of ZnO nanorods (NRs) have been attracted much attention for nanoscale sensors, detectors, and optoelectronics devices. Compared to the conventional etching methods, researchers have used micro- and nanopatterning of catalysts on which the direct deposition can be done site-selectively, by combining photochemical, catalytic, and self-assembled block copolymer tools (12-14). Line patterns of zinc oxide (ZnO) were site-selectively grown on the photo-patterned palladium catalysts through an electrode-less deposition process (15). However, there are no results available on the growth well-aligned nanorod arrays on patterned substrates.

Electron beam lithography (EBL) is one of the most versatile nanofabrication tools for making patterns for the growth of periodic ZnO nanostructures which has tremendous applications in opto-electronics, sensors, biomedicals and many other applications. However, the attachment of NRs on to the patterned surface is important of making 1D nanorod arrays. In case of electron beam lithography, the primary electron beam has either forward or backscatter as approached towards the resist coated substrates. The total electron yield (σ), which is the sum of the above depend solely upon the accelerating voltage that is used. Generally, higher accelerating voltage (σ <1) induce negative charges and the lower one (σ >1) makes it positively charged. As ZnO has negative charged carriers in majority, surface attachment supposed to be hindered at the higher voltage and conversely enhanced at the lower. Here we have demonstrated the selective growth of 1D nanorod arrays on the EBL patterned surfaces at lower energy. We have also demonstrated that how energy landscapes during EBL patterning plays a major role in self-assembly of ZnO NRs, using a seed layer instead of a catalyst.

Experimental Details

Al doped ZnO (AZO) seed layers of ~150 to 200 nm thick were deposited on the glass substrate using radio-frequency (RF) magnetron sputter at an ambient of 350^0C. Figure 1 shows the schematics of the fabrication of nanopatterns by EBL using PMMA, and the Scheme for the growth of aligned NR arrays selectively on EBL patterns. The AZO layer on glass was spin coated with ~50 nm of PMMA (Mol. wt. 950K) and post baked at 180 ^{0}C for 2 min. The samples were patterned using EBL at different beam energies of 20, 10, 5, and 2 keV.

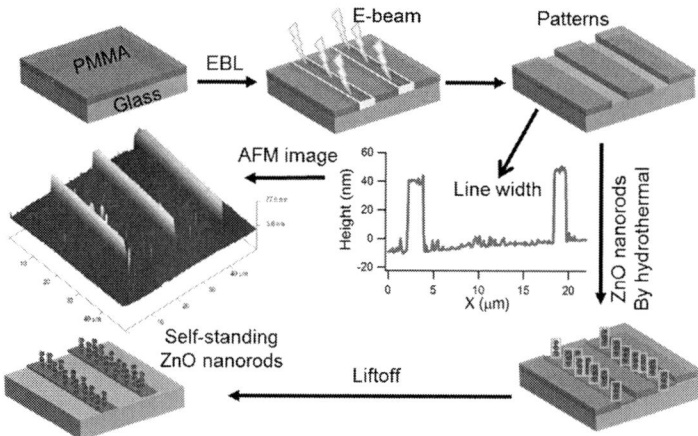

Figure 1: Schematics of fabrication of nanopatterns by EBL using PMMA, and the Scheme for the growth of aligned Nanorod arrays selectively on EBL patterns.

Figure 2 shows the typical EBL pattern at 2keV with varying pitch (from (a) to (c)) and with varying electron bean energy (from (d) to (f)). Upon electron beam irradiation, a layer of highly localized positive charges with a thickness about the mean escape depth (5–20 nm) is created right at the irradiated spot even though the net charge in the region is negative. The beam decelerates due to the accumulated negative charge in the whole region, and the emission yield increases to unity in the steady state (16), causing a negative shielding potential at μm distances away from the surface, which is not favorable for the attachment of the negatively charged molecules (17). The potential is positive close to the surface due to the presence of a positive charge layer. The surface charge can be manipulated by manupulating the beam energy as well as accelerating voltage as we have shown in Fig. 2 (d) to (f).

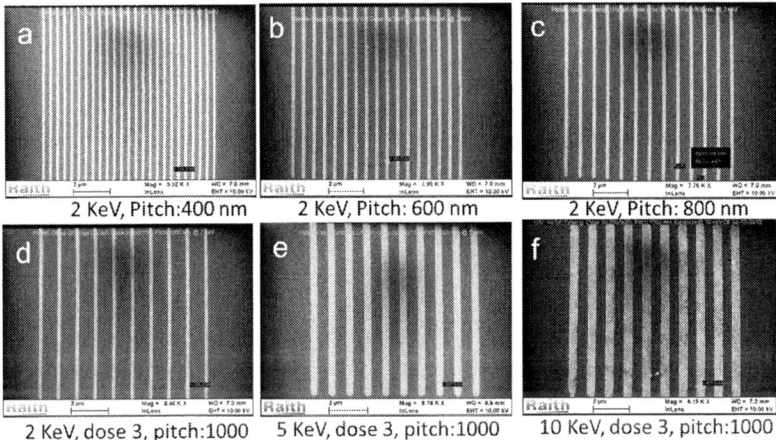

Figure 2: EBL patterns at 2 keV a - d with varying pitch, e at 5 keV and f at 10 keV.

After the patterning, the resist was undergone development in Methyl-isobutyl-ketone and isopropyl alcohol (MiBK+IPA) at a ratio of (1:3) for 30 sec, rinse in IPA for 15 sec then dried N_2. Patterned samples were processed for growth of ZnO nanorods using hydrothermal technique in a solution of using Zn $(NO_3)_2$ and hexamethylenetetramine (HMT) at 90^0C for 4 hrs. After successive ZnO growth, the unexposed PMMA were lifted-off using N-methyl-2-pyrrolidinone (NMP) for 10 min at 60^0C. Then samples were rinsed in DI and dried in nitrogen.

Results and discussion

Figure 3 shows the FE-SEM images of patterned surface at 20 keV, (a) Dose: 4.5, and (b) dose: 3, pitch 400 nm with processing for growth of ZnO nanorods using hydrothermal technique. Interestingly, ZnO nanorods did not grow on the surface although the patterned regions were very clear with visible AZO seed layers. As discussed earlier the beam decelerates due to the accumulated negative charge in the whole region at higher bean voltages, and the emission yield increases causing a negative shielding potential at μm distances away from the surface, which is not favorable for the attachment of the negatively charged ZnO nanorods. However, the situation changes as the beam voltages ramped down. Nanorod arrays start growing favorably on patterned surface at lower energy as shown here at 10 keV. Figure 4 shows the selective growth of nanorods on patterned areas for 10 keV, (a) dose: 4, pitch 400 nm, and (b) dose:6, pitch 400 nm. The nanorods are somehow organized at all dose values for this beam voltage. However, the growth of nanorod arrays becomes more organized on patterns formed even at lower energy of 5 keV.

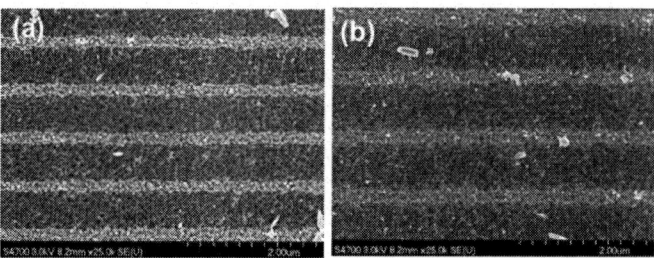

Figure 3. FE-SEM images of patterned surface at 20 keV, (a) Dose: 4.5, and (b) dose: 3, pitch 400 nm with processing for growth of ZnO nanorods using hydrothermal technique. NRs did not grow on the surface.

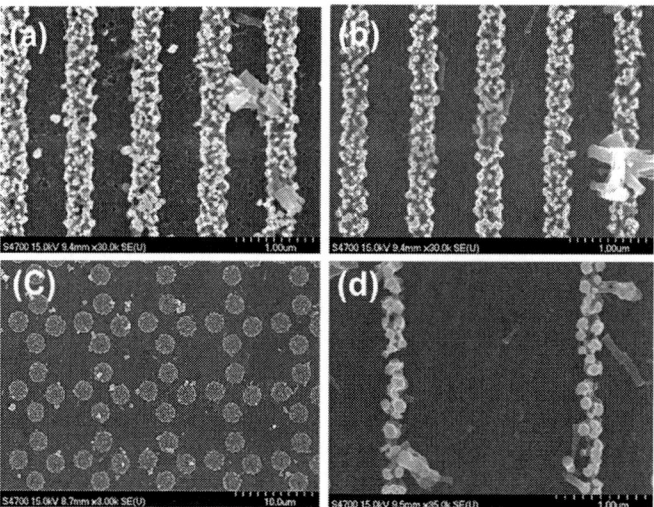

Figure 4. FE-SEM images of the selective growth of ZnO NRs nanorods on patterned areas for 10 keV, (a) dose: 4, pitch 400 nm, and (b) dose:6, pitch 400 nm.

Figure 5 shows the ZnO nanorod arrays at 5 keV for (a) dose:3, pitch 400, (b) dose:4.5, pitch 200, (c) dose:4.5, pitch 600, and (d) dose:4.5, pitch 100. The images show well-organized arrays of growth ZnO nanorods even at lower pitch values. Figure 6 shows the most organized growth patterns of ZnO NRs.

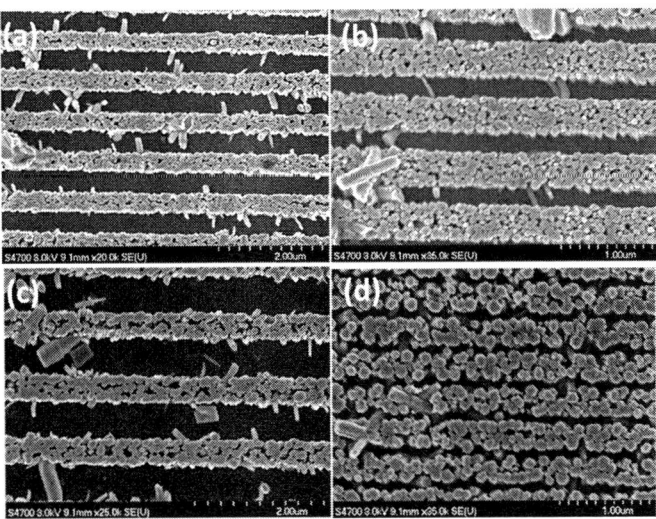

Figure 5. FE-SEM images of the selective growth of ZnO NRs nanorods on patterned areas for 5 keV, (a) dose:3, pitch 400, (b) dose:4.5, pitch 200, (c) dose:4.5, pitch 600, and (d) dose:4.5, pitch 100.

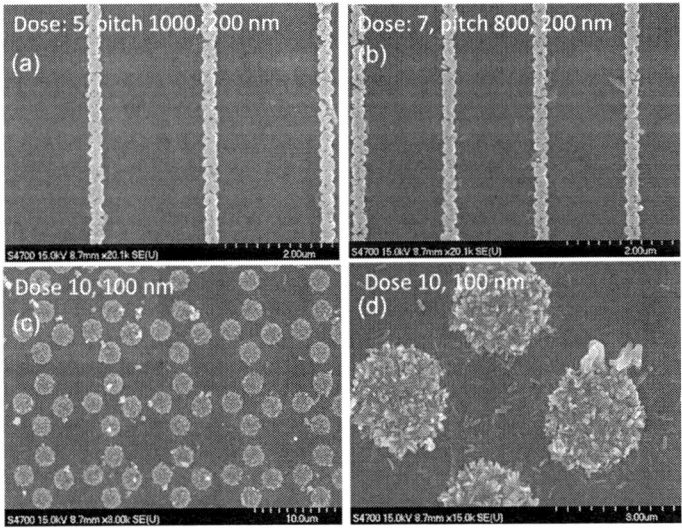

Figure 6. FE-SEM images of the selective growth of ZnO NRs nanorods on patterned areas for 2 keV, (a) dose:5, pitch 1000, (b) dose:7, pitch 800, (c) circle of 100 nm in diameter, dose:10, pitch 600, and (d) enlarged image of (c).

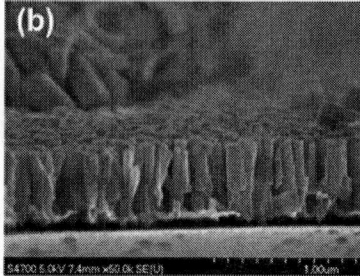

Figure 7. Side view of FE-SEM images of the selective growth of ZnO NRs nanorods on patterned areas for 2 keV for two magnification values.

Figure 7 shows the side view of the most distinctive growth patterns of well-aligned nanorod patterns for the beam energy of 2 keV. The nanorods have selectively grown on the patterned areas, irrespective of dose factor. The well-aligned nanorods of equal height (~ 500 nm) are clearly visible from the side view SEM images as shown in Fig. 7 (a) and (b) for the beam energy of 2 keV. Fine ZnO nanorods were site-selectively grown inside the pattern areas through low-temperature hydrothermal technique. These images clearly predict that the alignment of nanorod growth inside the pattern and probably their crystalline morphology are drastically affected by the beam energy responsible for pattern generation. The clean nanorod arrays with well-defined hexagonal facets and well-aligned columnar nature are only seen for the electron beam energy of 2 keV. Apart from that, the nanorods did not either bend or fall around due to better adherence to the seed layer.

The formation of ZnO nanorod arrays depends on the density of nucleation sites existing on the surface of the ZnO seed layer as well as underlying surface energy. The relationship between stable nuclei density (N*), total nucleation site density (n_s) and the Gibbs free energy (ΔG*) is shown in eq. 1 [18] as:

$$N^* = n_s \, exp \, (-\Delta G^*/k_b T) \qquad [1]$$

The Nucleation rate (N*) can be then calculated from $N'=N^*A^*\omega$, where A^* is the critical area of nuclei and ω is the rate at which atoms impinge on the nuclei (18). The formation and density of such nucleation sites is related to the surface energies of the seed layer. The surface energy is related to the difference in interatomic energy between the atoms on the surface of the bulk crystal that have unpaired bonding sites. A thermodynamic driving force acts to reduce the unpaired bonding sites, or in another way to reduce the Gibbs free energy on the surface (18). In view of the above, the electron beam modifies the surface energy of the underlying seed layer during the lithographic process. At higher beam energy (~ 20 keV), the incident electron beam decelerates due to the accumulated negative charges that already built on the surface during the irradiation process. This causes a negative shielding potential close to the surface down to a few micron, which is completely unfavorable for the attachment of the negative ZnO nanorod arrays. However, at comparatively lower beam energy, the secondary electrons are responsible for the pattern with the irradiation zone with a local positive charge (lower beam energy), which attracts the negatively charged ZnO ions to form nanorod arrays selectively in the pattern region only. We have recently shown that the surface potential on the ZnO seed layer depends on the annealing temperature, which modified the growth of ZnO NRs on the seed layer[19]. Therefore, we argue that the electron beam energy modifies the surface potential of the underlying seed layer, which eventually energetically favors the growth of NRs only at lower energy.

Figure 8 shows the transmission spectra of NRs on seed layer patterned substrates at various beam energy. Although there is no significant change in the band edge of all samples, there are some differences around 500 to 600 nm region. The transmission spectra for 2keV beam energy has characteristics of a Al:ZnO film with moderate thickness with a broad hump, which get reduced for 20 keV (no NRs were seen). However, the humps observed for 5 and 10 keV are very pronounced and believed to be related to the defect states in ZnO. This is also related to the defects associated with high density of ZnO NRs arrays. Similar trend was also observed for the absorption spectra as shown in the inset of Fig. 8.

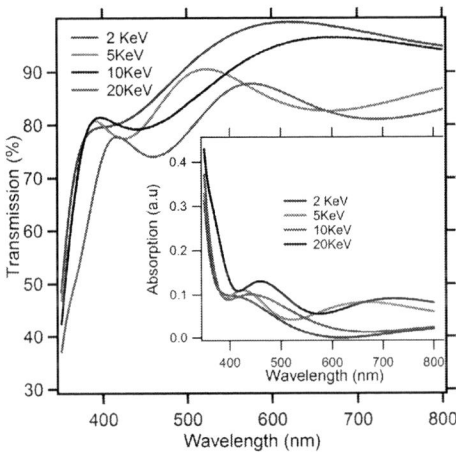

Figure 8. Transmission spectra of NRs/seed layer patterned substrates at various beam energy. The inset shows the absorption spectra.

Conclusion

In summary, we have demonstrated the attachment of highly aligned ZnO nanorod arrays on to the electron beam lithographically patterned surface at lower beam energy. At higher beam energy, the incident electron beam decelerates due to the accumulated negative charges that already built on the surface during electron beam irradiation process. This causes a negative shielding potential close to the surface at micron level. This becomes unfavorable for the attachment of the negative ZnO carriers. However, at comparatively lower beam energy (~ 5 keV or less), the secondary electrons are responsible for the pattern with the irradiation zone centered by the local positive field, facilitating growth of aligned nanorods. Hence, the negatively charged ZnO nanorod arrays can be grown at lower beam energy, which initiates site selective attachments. This protocol is very important for numerous applications in the field of sensors for biomedical as well as environments.

Acknowledgments

This work was supported by the NSF-CREST (CNBMD) Grant number HRD 1036494, NSF-MRI, and partially supported by DoD (CEAND) Grant Number W911NF-11-1-0209 (US Army Research Office).

References

1. J. Y. Cheng, C. A. Ross, H. I. Smith, E. L Thomas, *Adv. Mater.*, **18**, 2505 (2006).
2. Y. Cui, M.T. Bjork, J.A. Liddle, C. Sonnichsen, B. Boussert, A.P. Alivisatos, *Nano Lett.*, **4**, 1093 (2004).
3. L. Malaquin, T. Kraus, H. Schmid, E. Delamarche, and H. Wolf, *Langmuir*, **23**, 11513 (2007).
4. Y. Wang, D. Maspoch, S. Zou, G. C. Schatz, R. E. Smalley, and C. A. Mirkin, *Proc. Natl Acad. Sci.* USA, **103**, 2026 (2006).
5. L. Seemann, A. Stemmer, N. Naujoks, *Nano Lett.*, **7**, 3007 (2007).
6. Z. Fan, J.C. Ho, Z.A. Jacobson, R. Yerushalmi, R.L. Alley, H. Razavi, A. Javey, *Nano Lett.*, **8**, 20, (2008).
7. N. S. Losilla et al., *Nanotechnology*, **19**, 455308, (2008).
8. P. -Y. Chi, H.-Y. Lin, C.-H. Liu, C.-D. Chen, *Nanotechnology*, **17**, 4854 (2006).
9. S. Tanaka, M. Taniguchi, T. Kawai, *Jpn. J. Appl. Phys.*, **43**, 7346 (2004).
10. T. Djenizian, E. Balaur, P. Schmuki, *Nanotechnology*, **17**, 2004 (2006).
11. K. Sarveswaran, W. Hu, P. W. Huber, G. H. Bernstein, M. Lieberman, *Langmuir*, **22**, 11279 (2006).
12. N. Saito, H. Haneda, T. Sekiguchi, N. Ohashi, I. Sakaguchi, K. Koumoto, *Adv. Mater.*, **14**, 418 (2002).
13. S. Yamabi, H. Imai, *J. Mater. Chem.*, **12**, 3773 (2002).
14. Y. Masuda, W. S. Seo, K. Koumoto, *Langmuir*, **17**, 4876 (2001).
15. Lee Jae-Young, Yin Dehui, Horiuchi Shin, *Chem. Mater.*, **17**, 5498 (2005).
16. P. Pammer, R. Schlapak, M. Sonnleitner, A. Ebner, R. Zhu, P. Hinterdorfer, O. Hoglinger, H. G. Schindler, S. Howorka, *Chem. Phys. Chem.*, **6**, 900 (2005).
17. B. Saccà, C. M. Niemeyer, *Chem. Soc. Rev.*, **40**, 5910 (2011).
18. M. Ohring, *Materials science of thin films: deposition and structure.* (Academic Press: San Diego CA, 2002).
19. R. Mundle, T. Holloway, K. Zhang, M. Bahoura, A. K. Pradhan, *J. Nanosci. & Nanotechnol.*, **12** , 3938 (2012).

Micromolding of NiFe and Ni Thick Films for 3D Integration of MEMS

M. Cortes, J. Moulin, M. Couty, T. Peng, O. Garel, T. H. N. Dinh, Y. Zhu, M. Souadda, M. Woytasik, E. Lefeuvre

Institut d'Electronique Fondamentale, Université Paris Sud / CNRS, Orsay cedex 91450, FRANCE

Electrodeposition process was studied in order to obtain uniformly thick magnetic films which could be incorporated in magnetic microsensors and actuators. Ni and NiFe patterns of different dimensions were deposited by micromolding on a Cu/Ti seed layer. In the case of NiFe deposit, a linear deposition rate was calculated by measuring the deposit thickness using mechanical profilometry. The optimum pattern composition for the fabrication of magnetic microdevices was 80 at. % of Ni, that was found by comparing the magnetic properties of the deposits obtained from different conditions. The deposits structure was also characterized. All the parameters were studied as a function of the current density so that by controlling the electrodeposition conditions the characteristics of the deposits could be controlled. NiFe films were integrated in multilayered magneto-impedance sensors. They present good adherence even after performing annealing at both 300°C and 500°C and sensor sensitivity reaches 1%/Oe.

Introduction

The 3D integration of microsystems requires the stacking of patterns, which most of the times are obtained from different materials and in different shapes. For that reason it is very important to be able to control both the adhesion of the different layers, as well as thickness uniformity. For that purpose CMP (Chemical Mechanical Polishing) is generally used for etching and planarization of films in all kind of MEMS devices (1). In CMP process, planarization is achieved by combination of chemical and mechanical etching of the wafer, by using both a polishing pad and a corrosive and abrasive chemical slurry. However, due to the solid nature of the abrasives, this method presents several problems such as contamination of the samples, scratches or oxidation of the surface and induction of stress, which are particularly critical for magnetic materials.

This paper presents an alternative technique to CMP for depositing thick and flat patterns presenting a good adherence, in order to realize 3D magnetic and mechanical structures. Indeed, electrodeposition is an interesting alternative technique for getting stacked layers. Besides, thick films with low stress can only be deposited using electrodeposition, as it presents a high deposition rate and it is fully compatible with MEMS technology. Moreover, electrodeposition presents other advantages, such as easy control of the thickness and chemical composition, low cost, high simplicity of the experimental setup.

This work focusses on the micromolding of Ni and NiFe for the realisation of 3D magnetic and mechanical structures. The permalloy is indeed well known for its soft magnetic properties and can be integrated in both microdevices for energy storage or conversion, as a magnetic core, and as a functional soft magnetic film in actuators or sensors. The nickel present both ferromagnetic properties and mechanical hardness which can be useful for the realization of large membrane or moving parts of microdevices. Electrodeposition has been widely used for the preparation of magnetic patterns of different metals and alloys such as Ni, NiFe, CoNiFe (2-10). However in order to integrate them in a sensor or an actuator, it is sometimes necessary to apply different processes after the electrodeposition, e.g. annealing, and some of these process steps can lead to a lack of adhesion of the film on the seed layer. For that reason it is very important to find the optimum electrodeposition conditions which permit to obtain deposits with good adherence over the seed layer.

The aim of this work is thus to obtain a method for preparing uniformly thick magnetic films to be used in the fabrication of magnetic microsensors and actuators. In concrete, nickel micropatterns have been applied for the realization of piezoelectric transducers for energy harvesting which consist in 3D Ni structures with a copper sacrificial layer over glass/seed-layer. On the other hand, NiFe and Cu patterns have been piled up over Si/SiO_2/seed-layer for the realization of magneto-impedance (MI) sensors, which are made of a conductive Cu track surrounded by two ferromagnetic layers. In such devices, the ferromagnetic material is sensitive to an external field and the global impedance, measured by injecting a high-frequency current in the conductive track, varies with this external field through the skin depth. In this way, the softest ferromagnetic material has to be integrated, for increasing the sensitivity (11).

The deposits were characterized using both optical and mechanical profilometry, as well as EDS, XRD and AGFM measurements in order to study the deposition rate, composition, structure and magnetic properties as a function of the current density. Taking into account the final objectives, special attention was paid to thickness uniformity, adhesion on the seed layer, as these could compromise the viability of the device. Test consisting in the electrodeposition of a second layer of copper over Ni or NiFe patterns have also been done to complete the process study.

Materials and Methods

Sputtered Ti (10nm)/Cu (100nm) layers over 2 inches glass and Si/SiO_2 (100nm), thermically grown, were obtained by means of a Denton sputtering system. A thick AZ4562 photoresist layer up to 10 μm was spun and baked at 100°C for 50 s to remove the solvent; the photoresist was exposed to UV light and developed using AZ400K:H_2O (1:4) developer.

Electrodeposition has been carried out using a potentiostat/galvanostat Autolab PGSTAT302N with NOVA1.5 software in a two-electrode cell configuration at room temperature 20°C and without stirring. In order to prepare the Ni and NiFe deposits, solutions containing 0.75 mol/l $NiSO_4$•$6H_2O$, 0.02 mol/l $NiCl_2$•$6H_2O$, 0.4 mol/l H_3BO_3, 0.016 mol/l $C_7H_5NO_3S$ and 0.03 mol/l $FeSO_4$•$7H_2O$ (only in the case of NiFe) were used. Cu deposits were prepared galvanostatically with a solution containing 2 mol/l H_2SO_4, 0.3 mol/l $CuSO_4$•$5H_2O$, 1.4 10^{-3} mol/l Cl^- and 5 ml/l electrodeposit270 brightener. A copper plate placed at 5 cm from the working electrode was used as counter electrode.

For the MI sensor production, in order to test them and avoid the electrical short-circuit, it was necessary to etch the seed layer of the wafer. This was performed using a Roth & Rau IonSys 500 Ar+ ion beam etching.

A SEMFEG Phillips was used to study the morphology of the samples and a Hitachi 3600N SEM with EDS incorporated was used to analyze deposits composition. Films profiles were obtained using a DEKTAK 8 mechanical profilometer.

Structure of deposits was studied by means of a PANalytical X'Pert PRO MPD θ/θ Bragg-Brentano powder diffractometer of 240 mm of radius using Cu Kα radiation (λ = 1.5418 Å). Magnetic properties were characterized by means of an alternating gradient force magnetometer (AGFM) at room temperature.

Results and Discussion

Ni 3D structures

Fabrication and characterization. Patterns in shape of coils of some microns were obtained by applying a constant current density of J= -11 mA/cm^2 during t = 2 hours under stirring conditions, in order to obtain 20 µm-thick deposit. A glass/Ti/Cu layer electrode and a Ni plate were used as working and counter electrodes respectively with a distance between the electrodes of 1-2 cm. Ni deposits showing good adherence were obtained. However, these deposits showed a certain thickness non-uniformity of around 4-5 µm corresponding to a 20-25 % of variation on the whole wafer. In order to assure a more uniform thickness, the wafer was rotated 4 times every 30 min during electrodeposition. In order to obtain stacked patterns a structural and sacrificial Cu layer was electrodeposit over the Ni. Cu electrodeposition was performed on the whole wafer, without photoresist molds, at -20 mA/cm^2 untill a 20 µm thickness of copper covered the Ni patterns. In this case the working electrode was the previously electrodeposited Ni and the remaining seed layer. The wafer was rotated each 15 min to obtain a uniform thickness. After the Cu deposition, etching and polishing treatments were applied in order to remove the Cu on top of the Ni patterns and leave only the Cu in between (figure 1).

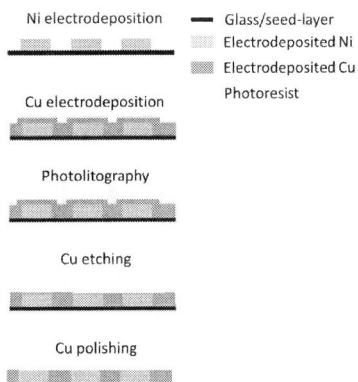

Figure 1. Schematic representation of the process to obtain one level of a 3D Ni structure.

All the steps were repeated several times to obtain a 3D structure. The final step consists in etching the Cu sacrificial layer in order to release the 3D Ni structure.

Discussion. The first Ni electrodeposited patterns presented well adherence over the Cu seed layer. However, when depositing the second Ni layer all the layers were detached from the seed layer because of the strain caused when growing Ni patterns on top of the others. In an attempt to improve the adherence of the structure a new seed layer of Ni/Cu was successfully tested as can be seen in figure 2 where one of these 3D structures is shown.

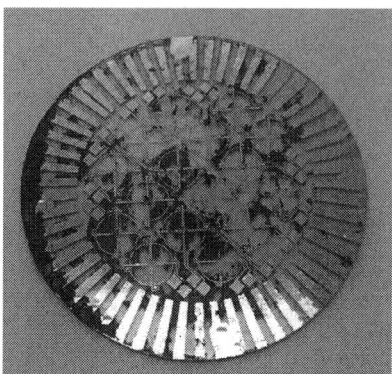

Figure 2. 3D Ni structure almost released after the etching of the Cu sacrificial layer.

Next work will consist in repeating ten or so times the process in order to fabricate a complex Ni structure that will be used as a piezoelectric actuator (12).

NiFe-based MI fabrication

Fabrication. NiFe electrodeposition was performed over 1x1 mm^2 test patterns under both direct and periodic two-step pulse current. Different conditions were tested in each case. For DC electrodeposition, current densities ranging from J= -5 to -40 mA/cm^2 and times from t = 5 to 40 minutes were used. In the case of pulsed deposition different sample sets were prepared consisting in;

PC1: J_c = -15 mA/cm^2, Q_c = 300 μC/cm^2, J_a = 5 – 45 mA/cm^2, Q_a = 100 μC/cm^2,
PC2: J_c = -10 mA/cm^2, Q_c = 200 μC/cm^2, J_a = 5 – 45 mA/cm^2, Q_a = 100 μC/cm^2,
PC3: J_c = -10 mA/cm^2, Q_c = 200 μC/cm^2, J_a = 2.5 – 22.5 mA/cm^2, Q_a = 50 μC/cm^2.

The total deposition time was kept to 30 min. Si/SiO$_2$/Ti/Cu substrates covered with patterned photoresist were selected as the working electrodes. The counter electrode was a Ni plate. The pulsed conditions were selected in order to obtain the optimum composition found previously applying DC, around 80 at.% Ni.

A second lithography step was performed to define new patterns on top of the already electrodeposited permalloy ones. These patterns were filled with electrodeposited Cu. A current density of -30 mA/cm^2 for times ranging from 5 to 20 min was applied in order to obtain a copper thickness between 2 and 12 μm.

Once the optimal conditions for the permalloy deposition were found, MI sensors with different width, length and thickness were prepared. The sensors consist in three stacked layers of different metals, two permalloy layers with a copper one in the middle, and also two copper contact pads at both sides. The fabrication difficulties lie in that sometimes some of the layers do not exhibit good adherence. Sensors of 5 mm long with a permalloy width ranging from 50-250 nm and a copper one from 10-100 nm have been performed by means of electrochemical techniques. Pulsed plated was used for permalloy layers and DC was chosen in the case of copper. By controlling the deposition time, the thickness of the deposits could be controlled. Different wafers were electrodeposited in order to test different permalloy and copper thickness as shown in table I. In figure 3 there are images of both a general view and a cross section of the sensors.

TABLE I. Permalloy and copper thickness for each wafer.

WAFERS	PM1 (µm)	Cu (µm)	PM2 (µm)	PMT (µm)
1	5.4	8-12	4.8	10.2
2	3.6	4.6-6	1.7	5.3
3	5.7	4.4-6	4.8	10.5
4	3.4	2.3-3	2.5	5.9

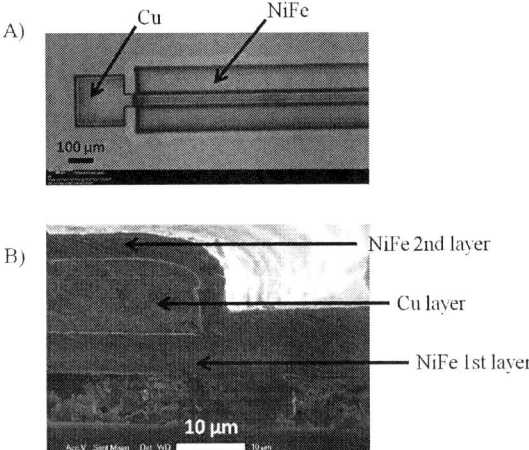

Figure 3. A) Optical image of a general view and B) SEM image of a transversal section of the sensors.

In order to validate the whole fabrication process of the sensor, the final steps have been performed. First, the Ti/Cu seed layer was removed by means of a ion beam etching to avoid short-circuit. The sensors were diced and annealed at both 300° C and 500° C. For MI optimization it is sometimes necessary to apply a transversal magnetic field. For that reason, in some cases, a magnetic field was applied at the same time that the annealing was performed. The sensors presented well adherence as they remained attached to the surface even after performing the annealing process and no changes were observed in the roughness or in the aspect of the metal layers.

Characterization. The nickel content (at.%) of the binary NiFe electrodeposits measured by EDS shows a nonlinear dependence with the current density (figure 4) which is in accordance with the literature (13,14). In the case of the pulsed plate set of samples the variations in the composition of the deposits were in the range of the equipment error so the tendency in composition variations was not observed.

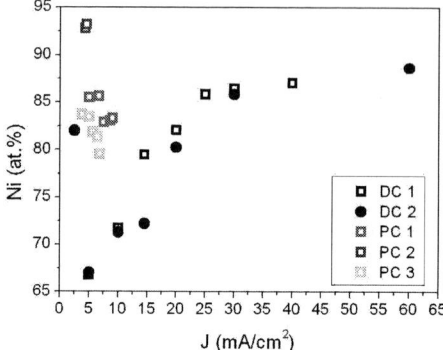

Figure 4. Representation of the deposits composition versus current density of patterns prepared under both DC and pulse plate conditions.

XRD analysis (figure 5) confirms that the deposits present a face-centered cubic (fcc) structure (Peaks identified from taenite, syn Iron-Nickel (JCPDS 47-1417)). A crystallite size around 9 nm was calculated with Scherrer equation [1], which is in agreement with the results of Cheung et al. (15):

$$\tau = \frac{K\lambda}{\beta \cos\theta}$$ [1]

Where K (shape factor) = 0.9 and β = FWHM (line broadening at half maximum). The calibration of the instrument, to subtract the FWHM caused by the equipment at each angle, was determined by the pattern LaB6 NIST SRM-660a.

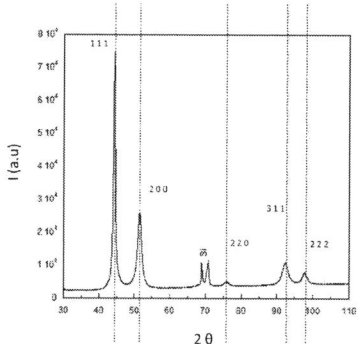

Figure 5. XRD of a 1 x 1 mm^2 test pattern obtained at J = -15 mA/cm^2 and a Ni atomic content of 80 %.

Magnetic measurements were performed. In figure 6 the relations between the relative permeability μr and the coercivity Hc versus the Ni at.% are shown.

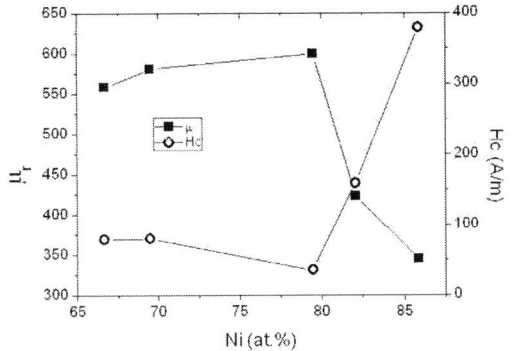

Figure 6. Relative permeability and coercivity versus Ni at.%.

Figure 7 shows two profiles of two different sensors. For these patterns the thickness differences are reduced from 30 % to 20 % compared with the once used for the initial tests.

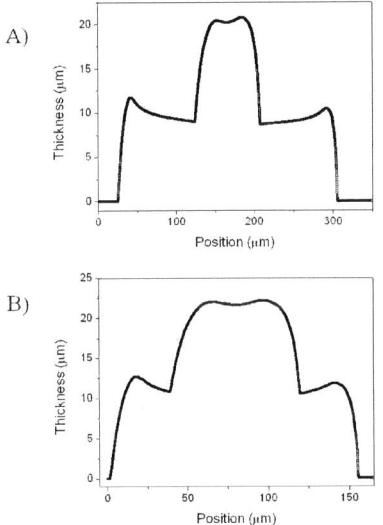

Figure 7. Mechanical profiles of A) type B and B) type C sensors from the wafer 2.

In order to test the viability of the sensors their sensibility was measured. Measurements were performed by measuring the resonance voltage generated by the (R,L) microsensor and an added capacitor. The sensor was excited at the optimal frequency, which consists in the one that gives the 50 % of the maximum resonance amplitude, while the supplied excitation current was fixed at 1 mA$_{rms}$ by using a 1 kΩ resistance in series. An external longitudinal DC magnetic field was created by a solenoid, which was controlled by a DC power supply. Finally the voltage measured at the terminals of the sensor, which is modulated by the DC field, was demodulated by a lock-in amplifier.

Discussion. Electrodeposition rates of each deposition conditions were calculated. In figure 8 the linear dependence of the deposition rate with current density for both DC and pulsed plate conditions is shown.

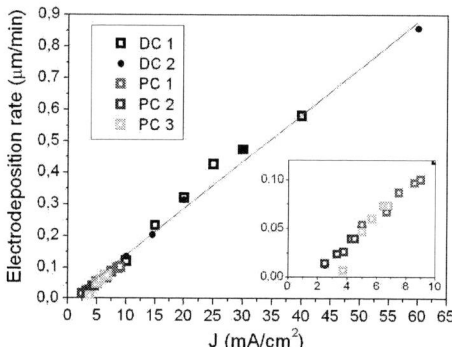

Figure 8. Deposition rate versus current density of patterns prepared under both DC and pulse plate conditions.

In the case of pulsed plate electrodeposition, the dependence was studied by taking the mean current density [2]:

$$\bar{J} = \frac{J_c t_c + J_a t_a}{t_a + t_c} \qquad [2]$$

The efficiency of the electrodeposition process was calculated by [3]:

$$\eta = \frac{n_{Ni} + n_{Fe}}{n_{Tot}} \qquad [3]$$

Where n are the number of moles in each case, which were calculated taking into account the thickness of the deposits and the density estimated from [4] (16), valid in the range of 50-100 % of Ni:

$$\rho = 8.22 + 0.68 \frac{\%Ni - 50}{50} \qquad [4]$$

The efficiency curves are represented in figure 9, the tendency of these curves is in good agreement with the one observed by Quemper et al (2), however the values in our case are higher, probably due to an overestimation of the alloy density as the porosity was not taken into account.

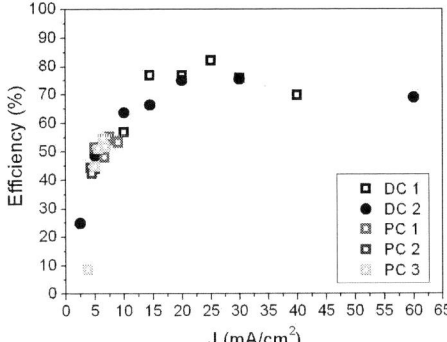

Figure 9. Efficiency of the electrodeposition process versus current density of patterns prepared under both DC and pulse plate conditions.

The correlation between the increase of Ni content (decrease of Fe) and the increase of efficiency observed in the case of the deposits performed under DC conditions is also observed by Costa (17) who found out that the larger the iron content in the deposits (the lower % of nickel) the smaller cathodic efficiency. That could be explained by the effect of the HER (hydrogen evolution reaction) that increases the pH near the electrode and this higher pH seems to favour Fe^{2+} reduction. The hypothesis posed by both Holkans (14) and Costa (17) after performing some studies about the role of the different constituents and parameters on electrodesposited NiFe was that Fe^{2+} discharge takes place through the hydroxide. For pulsed plate electrodeposition, the efficiency values were in good agreement with the ones obtained in DC conditions and the results were reproducible with former works (18,19). However, the global deposition rate (taking into account the dead times) was lowered by a factor of 3.

DC conditions lead to a non-uniform thick patterns, as can be seen in figure 10A. Asymmetry in the electrodeposition was observed not only within the same pattern but also in the different patterns of the wafer depending on their relative position. Positive/negative pulsed current which alternates deposition and etching can be used to overcome this problem. The pattern profiles in figure 10B show that the asymmetry in the deposit thickness was reduced by using the pulsed plate technique. However, overdeposition at the edges of the patterns, as well as non-uniformities between thicknesses of patterns from different parts of the wafer were still present. Thickness variations over the wafer were in the range of 30% and were characterized by a higher deposition rate on the top of the wafer, where the electrical contact was placed. This result could be explained by the hydrogen generation which might induce a vertical movement of the electrolyte. Horizontal positioning of the cathode in an upper position did not give better results, as hydrogen bubbles were trapped on the surface; however the thickness variations were reduced to a 5% when performing the deposition with the wafer in the vertical position and rotating it 3 times a quarter turn of the wafer each time.

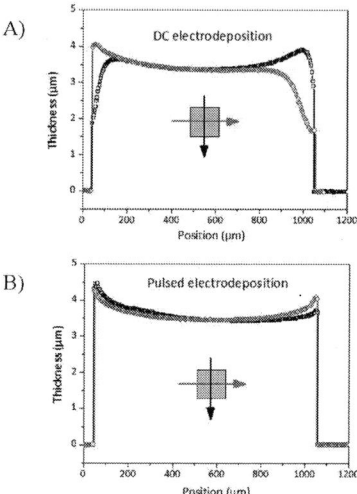

Figure 10. Profiles of a permalloy patter obtained A) in DC conditions. B) in pulse plate conditions ($\bar{j}$ = -7.5 mA/cm^2). These profiles have been measured along the two space directions in order to have an idea of the thickness homogeneity. The wafer contact was placed in the direction of the black arrow.

The optimum magnetic conditions i.e. the highest value of permeability, that it is well known that corresponds to the lowest value of coercivity (16), (H_c = 35 A/m, μ_r = 600), were attained for deposits with an 80 at.% of Ni. The magnetic polarization (P) decreases when the Ni percentage in the alloy increased (figure 11), which is in good agreement with the curves found in the literature (16). However, the values obtained in our case (P = 1,32 T for an 80 at.% of Ni) are higher than the ones found in the literature (P = 1,04 T for an 80 at.% of Ni).

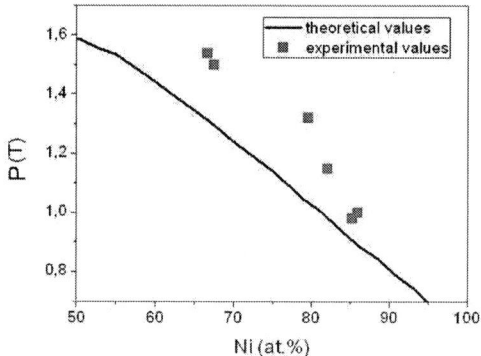

Figure 11. Theoretical and experimental curves of the variation of the magnetic polarization (P) with the Ni at.%.

This could be attributed to an overestimation of the density of the material, as using electrodeposition higher porosities than with other techniques can be commonly obtained. In figure 12 polarisation curves show that there were not significant changes in the magnetic properties of the deposits obtained by pulsed plate conditions compared with the ones obtained with DC conditions. From this results it can be said that permalloy deposits with 80 at.% Ni prepared by means of pulsed plate deposition may be suitable for MI sensors applications as they are more uniform in thickness.

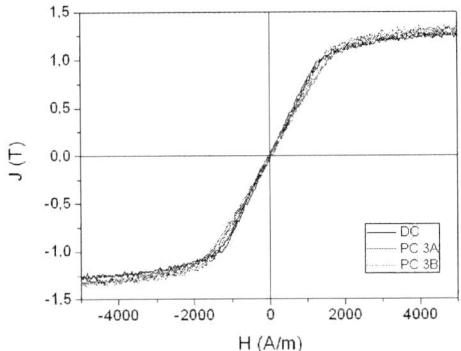

Figure 12. Magnetic polarization curves of deposits prepared under both DC (J = -15 mA/cm^2) and pulse plate conditions (PC 3A: J_c = -10 mA/cm^2 t_c = 20 ms, J_a = 22.5 mA/cm^2 t_a = 2.22 ms; PC 3B: J_c = -10 mA/cm^2 t_c = 20 ms, J_a = 15 mA/cm^2 t_a = 3.33 mn.

The sensors voltage measurements results were compared with former studies carried out with sputtered FeCuNbSiB (Finemet®) deposits as magnetic layers instead of permalloy ones which were annealed at 300°C with a transversal magnetic field (20). The sensors sizes were 5 mm length, 250 µm wide for the permalloy or the finemet and 100 µm wide for the copper. In the case of NiFe-based MI sensors, the thicknesses were 5.4 µm for the first layer of permalloy, 10 µm for copper and 4.8 µm for the second layer of permalloy. For the Finemet-based sensors all the layers were 500 nm thick.

The sensors tested showed variations of 0.012%/Am^{-1} (1%/Oe) and 0.0028/Am^{-1} (0.22%/Oe) around H = 0 for non-annealed and annealed NiFe-based sensors. Variations for Finemet-based sensor were -0.004%/Am^{-1} (0.32%/Oe) around H=1600 A/m. This demonstrates that there is no need to apply an annealing to the NiFe-based sensors in order to reach high sensitivity. In addition, there is no need for DC magnetic field polarization to reach this sensitivity. Finally, NiFe-based sensors presented a lower hysteresis than the Finemet-based ones (figure 13).

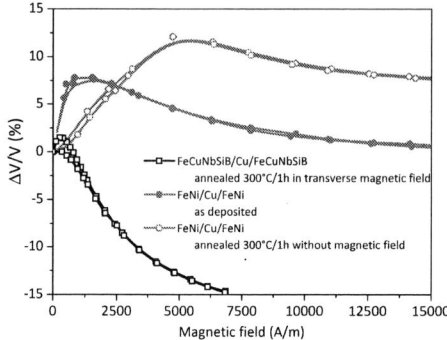

Figure 13. Relative value of the voltage with the magnetic field for two permalloy-based sensors (annealed and non-annealed) and a finemet-based sensor.

Conclusions

In a first step the study of Ni and NiFe deposition in the shape of different patterns was performed in order to find the optimum electrodeposition conditions and seed layers to obtain well adhered, uniform. In the case of Ni the best results were obtained for a Ti/Ni/Cu seed layer and a constant current density of $J = -11mA/cm^2$, while for permalloy deposits a Ti/Cu seed layer and pulsed plated conditions were used. The thickness uniformity of the layers was optimized either by rotation the wafer during the electrodeposition or by adjusting the pulsed conditions. Once electrodeposition conditions determined from test patterns presented the desired adherence, composition, structure, thickness and magnetic properties, they have been used and permitted the preparation of both Ni and NiFe structures for their application in sensors and actuators. NiFe-based sensors showed higher sensitivity than Finemet-based one without need of either annealing or DC field polarization.

Acknowledgments

This work has been financially supported by the Labex LaSIPS and by the Conseil Général de l'Essonne through the use of the equipments of the Centrale de Technologie Universitaire IEF-MINERVE.

References

1. C. Kourouklis, *Sens. Act. A*, **106**, 263 (2003).
2. J.M. Quemper, S. Nicolas, J.P. Gilles, J.P. Grandchamp, A. Bosseboeuf, T. Bourouina, E. Dufour-Gergam, *Sens. Act. A*, **74**(1–3), 1 (1999).
3. S. Liao, *IEEE Trans. Mag.*, **23**(5), 2981 (1987).
4. I. W. Wolf, *J. Appl. Phys.*, **33**(3), 1152 (1962).
5. Y. Sverdlov, Y. Rosenberg, Y.I. Rozenberg, R. Zmood, R. Erlich, S. Natan, Y. Shacham-Diamand, *Microelectron. Eng.*, **76**(1–4), 258 (2004).

6. X. Liu, M. Shamsuzzoha, G. Zangari, *J. Electrochem. Soc.*, **150**(3), C 159 (2003).
7. I. Tabakovic, V. Inturi, J. Thurn, M. Kief, *Electrochim. Acta*, **55**, 6749 (2010).
8. L. Péter, A. Csik, K. Vad, E. Tóth-Kádár, Á. Pekker, G. Molnár, *Electrochim. Acta*, **55**, 4734 (2010).
9. E.V. Khomenko, E.E. Shalyguina, N.G. Chechenina, *J. Magn. Magn. Mater.*, **316**, 451 (2007).
10. J. Li, Z. Zhang, J.Y. Yin, W. Geng-hua, C. Cai , J.Q. Zhang, *Trans. Nonferrous Met. SOC. China*, **16**, 659 (2006).
11. L.V. Panina, K. Mohri, *Sens. Act.* A, **81**, 71 (2000).
12. M. Deterre, E. Lefeuvre, Y. Zhu, M. Woytasik, A. Bosseboeuf, B. Boutaud, R. Dal Molin, *IEEE MEMS*, **1**, 249 (2013).
13. B. Robotin, A. Ispas, V. Coman, A. Bund, P. Ilea, Waste Manage., *Waste Manage.*, **33**, 2381 (2013).
14. J. Horkans, *Plating Parameters*, **128**(1), 45 (1981).
15. C. Cheung, G. Palumbo, U. Erb, *Scripta Metall. Mater.*, **31**(6), 735 (1994).
16. R.M. Bozorth, *Ferromagnetism*, p.280, D.VAN Nostrand Company INC, Canada (1951)
17. V. Costa, *Surf. Coat. Tech.*, **96**, 135 (1997).
18. S. Roy, A.Connell, M.Ludwig, N.Wang, T.O'Donnell, M.Brunet, P.McCloskey, C. OMathuna, A.Barman, R.J. Hicken, *J. Magn. Magn. Mater.*, **290**, 1524 (2005).
19. F. Giro, K. Bedner, C. Dhum, J. E. Hoffmann, S. P. Heussler, L. Jian, U. Kirsch, H. O. Moser, M. Saumer, *Microsyst. Technol.*, **14**, 1111 (2008).
20. T. Peng, J. Moulin, Y. Le-Bihan and F. Alves, *Nanosensors, Biosensors, and Info-Tech Sensors and Systems* Proc. SPIE 8691 (2013)

Sm-Co Thick Films Micromolding

J. Moulin, M. Woytasik, D. Belghiti, K. Chouarbi

IEF, UMR 8622, Univ. Paris Sud / CNRS, 15 rue G. Clémenceau, 91405 Orsay, France

This paper presents a study of the scale effect occurring during electrodeposition in the (micro)molding process of a rare earth - transition metal alloy : SmCo. Different cathode sizes have been implemented in the electrodeposition set up: a large strip (10 cm x 1.5 cm), a matrix of 1 mm² square and matrix of 5 – 50 μm micropatterns. Results concerning composition, deposition rate, adherence and thickness uniformity of the patterns show an influence of the scale effect: the decreasing of the conductive surface of the cathode leads to thinner films containing less samarium, but also a lower oxygen content. Patterns are then free of cracks, more uniform in thickness and present a good adherence to the copper seed layer.

The influence of the deposition time has been also studied for micromoulding process, showing an increase of the samarium content but a decrease of the deposition rate during the deposition.

Introduction

The need for micropatterning of permanent magnets increases since magnetic actuation and detection emerge in MEMS world. A promising domain is for instance the integration of several functions in a lab-on-chip, like move control, numbering and sorting of magnetic nanoparticles. However, despite promising results found by triode sputtering or composite material preparation, fabrication and patterning of powerful micromagnets is still an issue. Indeed, triode sputtering with high deposition rate allowed depositing NdFeB and SmCo thick films (1), but the high temperature needed during deposition limits the integration. At the other hand, micromagnets have been realized by bonding wax and NdFeB powder (2), by including SmCo and NdFeB particles in PDMS or other polymer (3-7) or by ink printing (8). But these techniques still lead to a limited flux density due to the weak volume ratio of magnetic material

An alternative way would be micromolding, i.e. electrodeposition in a thick photoresist mold. Electrodeposition of Sm-Co has been reported in the past years (9-12), showing potentialities in the aimed application. However, it has been found that both oxygen contamination, stress and hydrogen adsorption are responsible for weak adhesion on the seed layer and cracks in the films which limit film thickness and magnetic properties. We present a study of Sm-Co micromolding, leading to the realization of microdots with a high aspect ratio. The composition, the deposition rate and the integrity of the film have been studied as a function of the dot size and results reveal an influence of the scale effect.

Experimental procedure

Two electrolytic solutions with different concentrations of Co(II) ions were studied. Both contain 1 mol.l^{-1} of samarium sulfamate (Sm(NH$_2$SO$_3$)$_3$) obtained by mixing sulfalmic acid and samarium oxide. One solution contains 0.03 mol.l^{-1} of cobalt sulfamate (Co(NH$_2$SO$_3$)$_2$) and 0.09 mol.l^{-1} of glycine (NH$_2$CH$_2$COOH) and the other contains 0.06 mol.l^{-1} of cobalt sulfamate and 0.18 mol.l^{-1} of glycine.

The process related to deposition in a Hull cell has been detailed in (13,14). Concerning micromolding, the films were electrodeposited through photoresist molds on a Cu (100 nm)/Cr (10 nm) seed layer sputtered on a SiO$_2$ (100 nm)/Si substrate. The molds were realized by UV photolithography using a 20 μm thick AZ4562 photoresist. The resist is classically spincoated and then pre-baked on a hot plate at a final temperature of 90°C (initial temperature 20°C) for 1 h. It is exposed in a EVG®620 mask alignment system with a 350 mJ.cm^{-2} energy dose. The development is realized in a 1:4 diluted alkaline solution (AZ 400 K) for 5 min. Then the electrodeposition is performed in a common two-electrodes set-up, a platinum wire used as counter electrode (see Fig. 1). The current density is in the 50 – 100 mA/cm^{-2} range. The deposition time was varying, from 5 to 30 minutes. The bath temperature was 20°C and the solutions were not stirred during deposition. After deposition, the samples were immediately removed and rinsed with deionized water and dried with nitrogen.

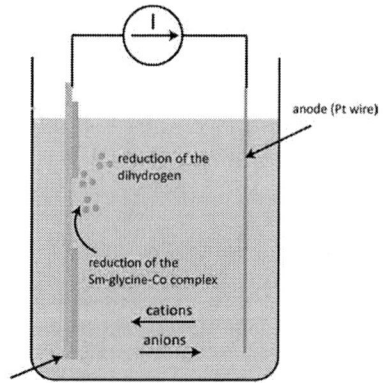

Figure 1. Experimental set-up for micromolding

The thickness of the films was measured both by SEM and by using a Dektak mechanical profilometer. The sample composition was characterized using a Hitachi 3600N SEM coupled with an Energy-Dispersive X-ray spectroscopy (EDS). The samarium, oxygen and cobalt elements have been quantified using the L, K and K peaks respectively, with a 10keV electron beam. In these conditions, following Monte Carlo simulations, the corresponding analyzed depth is in the range of 250 nm, what limits the surface contribution to the analysis.

Magnetic characteristics have been extracted from hysteresis loops drawn at room temperature using a Princeton Measurement Alternating Gradient Field magnetometer (AGFM).

Results and discussion

Composition and deposition rate

Former studies in a Hull cell, using a copper cathode, allowed validating the process and the electrolytic solutions [13,14]. The size of the deposit was 10 cm x 1.5 cm, after covering a part of the cathode with photoresist. The influence of the elaboration parameters on the film composition has been roughly defined. In particular, it was showed that the continuous current supply only can lead to samarium contents larger than 7% in the films. In addition, 1 mm x 1 mm square patterns were realized using thin S1818 photoresist (which thickness is 1.8 µm) for molding test on Si substrates metalized by Cu(100 nm)/Cr(10 nm). These samples allowed also magnetic characterisations using an Alternating Gradient Field Magnetometer. Finally, micromolding with a high aspect ratio has been performed using thick AZ4562 resist. Different patterns were available, from 5 µm in diameter dots to 50 µm x 50 µm squares.

The deposition in DC mode highly depends on the diffusion properties of the different ions in the electrolyte, thus on the respective sizes of the cathode and the bath, favoring diffusion in 1D (comparable sizes of the cathode and the bath, case of Hull cell) or 2D (small cathode compared to the bath, case of micromolding). Thus the deposition has been studied as a function of the different cathode sizes in order to determine the scale effect on the composition and the deposition rate.

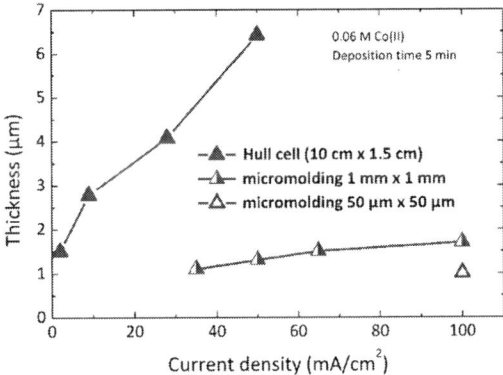

Figure 2. Film thickness as a function of the current density for different pattern dimensions

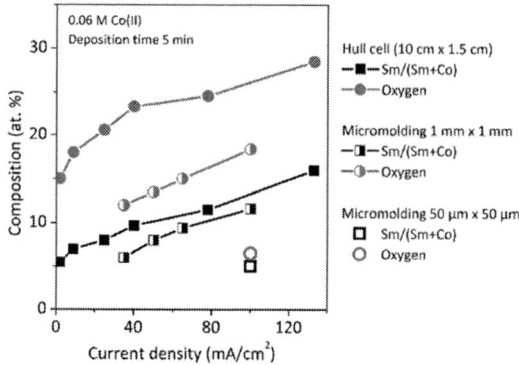

Figure 3. Film composition as a function of the current density for different pattern dimensions

The results are presented in Fig. 2 and 3. The dependence of the deposition rate and the composition with the current density is common to all deposit, independently on the scale: the deposition rate, the Sm/(Sm/Co) ratio and the oxygen contamination increase with the current density. However, the results show that these parameters are linked to the dimensions of the pattern. Indeed, in order to reach in the microdeposit a composition and a deposition rate similar to the ones in the Hull cell, a much higher current density has to be applied. It is thus supposed that the scale factor impacts the ions diffusion from the electrolyte to the cathode surface. The consequence is a variation in the cobalt and samarium relative deposition rates leading to a variation in the composition of the film. As shown in Fig. 3, the deposition rate is equivalent for films with similar composition but deposited on different cathode size. This means that the deposition rate is mostly driven by the composition and not by the scale effect. In addition to diffusion, another factor influencing the composition and linked to the scale effect is the pH increase close to the cathode surface. This increase in pH, due to the HER, is less pronounced with small patterns than with larger ones (as in Hull cell). Thus the oxygen contamination of the film through incorporation of $Sm(OH)_3$ is limited using micromolding.

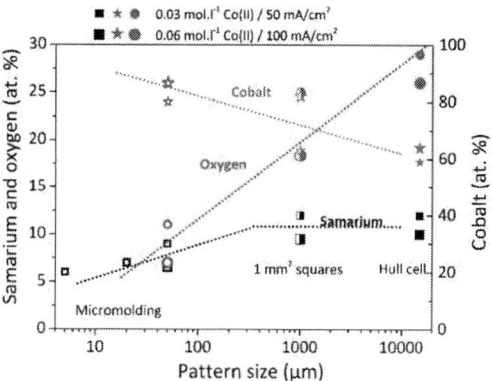

Figure 4. Scale effect on the composition

Fig. 4 presents the composition of films deposited using the two solutions as a function of the pattern size: ~1.5 cm for Hull cell, 1 mm for molding and 50 µm for micromolding. The scale effect shows that both cobalt and oxygen contents are approximately varying with the logarithm of the effective cathode surface. However, the oxygen content increases with the pattern size while the cobalt content decreases. The samarium contents also increases with the pattern size, but presents a limit around 10-12 %. This could be linked to the Sm electrodeposition mechanism which is an induced codeposition thanks to the cobalt presence [9]. Indeed, at a given current density, the relative Co(II) consumption is higher near a large cathode surface, limiting the codeposition phenomenon and promoting the solvent reaction, thus the pH increase. In addition to these measurements, the Sm/(Sm+Co) ratio has been measured for 5 and 20 µm microdots and results are summarized in Fig. 5 (the oxygen content in such microdots was imprecise). The micromolding clearly induces a diminution of the relative content in samarium. One hypothesis could be that the diffusion of samarium and cobalt are not equally affected by the scale effect, as their sizes and concentrations in the solution are very different.

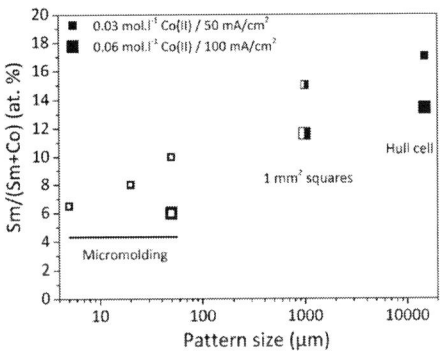

Figure 5. Scale effect on the Sm/(Sm+Co) ratio

Indeed, in case of samarium which is at high concentration in the solution, the diffusion is not affected by the cathode size. At the other hand, in case of cobalt which is at smaller concentrations, the diffusion adopts a spherical 2D configuration and by considering a circular cathode which radius is r_0, the ions flux per unit area is [15]:

$$J_s(t) = D\left(C_0^{sol} - C_0^{cath}\right)\left[\frac{1}{\sqrt{\pi Dt}} + \frac{1}{r_0}\right] \qquad [1]$$

with D the diffusion coefficient, C_0^{sol} and C_0^{cath} the initial concentration of ions in the solution and at the cathode. As the process is realized in DC mode with long deposition times, the transient stage can be neglected:

$$J_s(t) \cong \frac{D\left(C_0^{sol} - C_0^{cath}\right)}{r_0} \qquad [2]$$

This result shows that the cobalt flux decreases while the cathode size increases, what explains an increasing of the Sm/(Sm+Co) ratio.

The interesting point is that by comparison with Hull cell and millimetre-sized molds, micromolding leads to lower samarium content at a given current density, but also to lower oxygen contamination for a given samarium content. Using the solution containing 0.06 mol.l^{-1} of Co(II), a pattern containing 5 % of Sm contains also 6% in oxygen in shape of a 50 μm microdot, 12 % in shape of a 1 mm square and 15 % in the Hull cell deposit.

Influence of deposition time

The influence of the deposition time has also been studied in 50 μm x 50 μm squares microdots (see Fig. 6). The film thickness increases with time but the deposition rate decreases from 0.3 μm/min at 5 min to 0.1 μm/min after 30 minutes of deposition. This result could be associated to a decrease of the efficiency during film growth. Inversely, the relative quality of the film increases with deposition time: Sm/(Sm+Co) ratio increases from 5 to 10 % while oxygen contamination increases only from 6.5 to 8.5 %. This evolution should not be due to the measurement technique that analyses less than 500 nm below the film surface. In addition, this result is consistent with the scale effect previously presented and showing that Sm/(Sm+Co) ratio, oxygen and deposition rate are linked (see Fig. 2). As a conclusion, the process of incorporation of samarium in the film is favoured while deposition speed decreases at a given current density. This can be achieved by using long deposition time in small patterns.

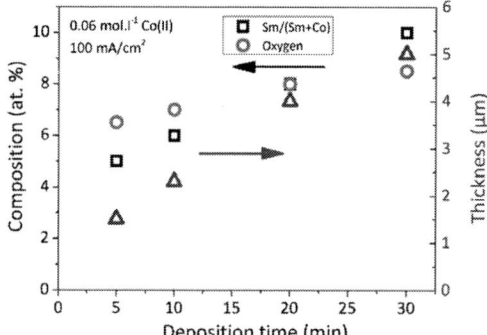

Figure 6. Composition and film thickness as a function of deposition time for 50 μm x 50 μm squares microdots

Film adherence

The scale effect can also affect film adherence. Indeed, the large stress in Sm-Co induced by the high deposition rate and the hydrogen generation is generally responsible for cracks and peeling of the films [10, 12]. As showed on Fig. 7 (left), large patterns peel from the seed layer. However, SEM observations show that the film adherence and its integrity were ameliorated by micropatterning, as patterns smaller than 30 μm in diameter are free of defects (Fig. 7, right).

Micromolding of 5 μm dots has been achieved with a high yield (see Fig. 8, left). An accidental fall of one dot shows that the copper seed layer has been torn but the dot remained attached to the seed layer (Fig. 8, right). This reflects the good adherence properties of the pattern processed using micromolding. As a final result, dots with a diameter of 5 μm and a thickness of 8 μm (i.e. an aspect ratio of 1.5) have been processed.

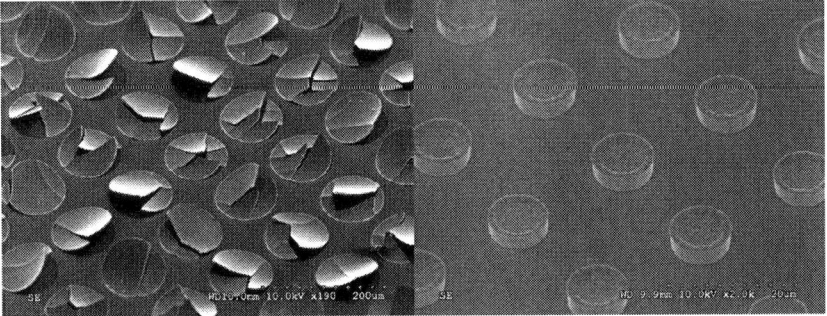

Figure 7. Micromolding of 100 μm (left) and 10 μm (right) patterns

Figure 8. Micromolding of 5 μm in diameter dots

Film thickness uniformity

Finally, it has been showed that micropatterning also ameliorates the thickness uniformity. The Fig. 9 compares Hull cell deposition (perpendicularly to the gradient of the current density), 1 mm x 1 mm molding and micromolding. The results show that the uniformity in current density over the opening, thus the film composition and thickness, is better for microscopic openings than for macroscopic ones. This important result is linked to the increase of the current density close to non-conductive parts of the cathode [16]. It leads to a better integration ability of the film in case of stacking and allows a better uniformity of the magnetic field generated by the magnet.

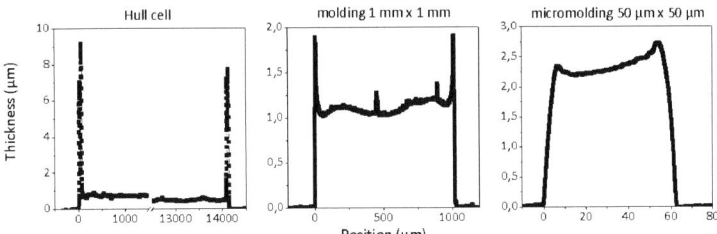

Figure 9. Scale effect on the thickness uniformity

Magnetic properties

The magnetic properties of 1 mm x 1 mm samples have been characterized using an AGFM. The hysteresis loops shows the highest in-plane coercivity for 5 min deposition at 100 mA/cm^2 using the solution containing 0.06 mol.l^{-1} of Co(II). The coercive value is 13.5 kA/m and the square ratio is 0.42. With the solution containing 0.03 mol.l^{-1} of Co(II), the highest coercive field obtained is 8,8 kA/m and corresponds to 5 min of deposition at a current density of 50 mA/cm^2. Finally, the coercive field of 50 μm x 50 μm patterns deposited during 30 min at 100 mA/cm^2 using the solution containing 0.06 mol.l^{-1} in Co(II) solution a is 3.2 kA/m. The optimization of both the compositions, in particular the samatium content and the structure that can be only achieved by annealing, will allow obtaining larger coercivities in future works.

Conclusion

Micromolding of Sm-Co dots has been achieved using electrodeposition on a copper seed layer. The comparison between deposition in a Hull cell, molding of 1 mm squares and micromoding of 5-50 μm patterns shows that the scale effect has an influence on both the composition and the deposition rate of the film. Micromolding leads to thinner films containing less samarium but free of cracks, more uniform in thickness and presenting a good adherence to the copper seed layer. The highest coercive field obtained is 13.5 kA/m. Next works will consist in optimizing the elaboration conditions, in particular annealing temperature, in order to form the magnetic SmCo$_5$ and Sm$_2$Co$_{17}$ phases.

Acknowledgments

This work has been financially supported by the French Agence Nationale pour la Recherche, through the Project SAIPON and by the Conseil Général de l'Essonne through the use of the equipments of the Centrale de Technologie Universitaire IEF-MINERVE.

References

1. A.Walther, C. Marcoux, B. Desloges, R. Grechishkin, D. Givord, N.M. Dempsey *J. Magn. Magn. Mat.* **321**, 590 (2009)
2. S.-S. Je, N. Wang, H.C. Brown, D. P. Arnold, J.Chae, *Proc. of Transducer 2009 conference (Denver, 21-25 june 2009)*, 885-888
3. N. Weber, D. Hertkorn, H. Zappe, A. Seifert, J. Microelectromech. Syst. **21** (5), 1098 (2012)
4. A. Khosla and Bonnie L. Gray, *ECS Trans.* **45**(3), 477 (2012)
5. A. Khosla, J.L. Korčok, B.L. Gray, D.B. Leznoff, J.W. Herchenroeder, D. Miller, Z. Chen, *Proc. of the SPIE conf. Microfluidics, BioMEMS, and Medical Microsystems VIII*, **7593** (2010).
6. M. Pallapa and J. T. W. Yeow, *J. Electrochem. Soc.*, **161**(2), B3006 (2014)
7. D. D. Hilbich, A. Khosla, B. L. Gray and L. Shannon, *Proc. of the SPIE conf. Microfluidics, BioMEMS, and Medical Microsystems X*, **7929** (2011)

8. S. Schwarzer, B. Pawlowski, A. Rahmig and J. Töpfer, *J. Mat. Science*, **15** (3), 165 (2004)
9. J. C. Wei, M. Schwartz and K. Nobe, *J. Electrochem. Soc.* **155**(10) D660 (2008)
10. M. Schwartz, N. V. Myung and K. Nobe, *J. Electrochem. Soc.* **151**(7),C468 (2004)
11. J. C. Wei, M. Schwartz and K. Nobe, *ECS Trans.* **16**(45), 129 (2009)
12. J. Chen and L. Rissing, *Journal of applied Physics* **109**, 766 (2011)
13. K. Chouarbi, M. Woytasik, E. Lefeuvre and J. Moulin, *Microsystem Technologies*, DOI 10.1007/s00542-013-1806-z (2013).
14. K. Chouarbi, E. Dufour-Gergam, M. Woytasik, E. Lefeuvre and J. Moulin, *J. Electrochem. Soc.* **159**(10), D592 (2012).
15. Fundamentals and Applications, 165, Wiley (2001)
16. A. C. West, M. Matlosz and D. Landolt, *J. Electrochem. Soc.*, **138**(3), 728 (1991)

64

Design and Modeling of a Novel Two Dimensional Nano-Scaled Force Sensor Based On Silicon Photonic Crystal

[a]Tianlong Li, [a, b]Longqiu Li*, [a]Wenping Song, [a]Guangyu Zhang and [c]Yao Li*

[a] School of Mechatronics Engineering, Harbin Institute of Technology, Harbin, China

[b] Postdoctoral Station of Material Science and Engineering, Harbin Institute of Technology, Harbin, China

[c] Center for Composite Materials and Structure, Harbin Institute of Technology, Harbin, China

*longqiuli@gmail.com

*yaoli@hit.edu.cn

Photonic crystal, which is an attractive optical structure for controlling and manipulating the flow of light, has been widely used to design mechanical sensors in microelectromechanical systems (MEMS) and nanoelectromechanical systems (NEMS). In this work, a novel two dimension nano-scaled force sensor based on silicon photonic crystal, in which a nanocavity is embedded in an L-shaped microcantilever, is developed and studied numerically. This microcantilever is extremely sensitive to the small refractive index changes produced by two dimensional component forces. The relationship between the force and the output wavelength is determined using finite element method and finite difference time-domain method. As we found, the range of the force sensor in each component force in X and Y directions are 0-1μN.And the resolutions of each component force in X and Y directions are 1.891 nm/μN and 1.418 nm/μN, respectively, for a 30μm long and 15μm wide cantilever. The novel photonic crystal sensor shows promising linear characteristics as an optical nanomechanical sensor.

Introduction

With the advantages of ultracompact size, high contrast and easy integration, photonic crystal, which is composed of periodic nanostructures and can be used to fabricate ultra-compact and highly wavelength selective optoelectronic devices, is one of the most promising platform in MEMS and NEMS [1, 2]. The size of most optical components can be greatly reduced and the strength of light-matter interaction can be significantly increased as a result of the extreme properties of photonic crystals, such as optical add-drop multiplexers [3], wavelength selective optical switches [4], tunable couplers [5] and tunable ring resonators [6]. Various micro-displacement sensors based on photonic crystal have been proposed in the past few years [7-9]. Xu et al. [8] design a micro displacement sensor based on line-defect resonant cavity in photonic crystal. The line-defect resonant cavity consists of a fixed photonic crystal segment and a moving one. As they found, the shift of output light intensity is a linear function of the distance between the fixed and mobile segments. Stomeo et al. [10] fabricate the force sensors based on photonic crystal nanocavity. The output wavelength is sensitive to the length shift of the nanocavity along the longitudinal direction. The range is between 0.25Gpa and 5Gpa, and the achieve pressure resolution is 5.82nm/GPa, with a sensitivity in the milli-newton range. Although this force sensor has a wide range and a high resolution, the force has to be loaded on the top of nanocavity and the accuracy of force sensor is sensitive to its loading location. It is very difficult to determine the loading location exactly for its application. Hence this sensor is extremely hard to be used MEMS and NEMS. However, the bending of microcantilever is not sensitive to the load location. The application point is only need to locate on one side of cantilever. Therefore, combining the nanocavity with the microcantilever can reduce the effect of loading location on measurement accuracy.

The microcantilevers have also been studied in the field of biosensors recently [11-20]. The bending of the microcantilever is extremely sensitive to changes loading on the surface of the microcantilever. Atomic Force Microscope (AFM) is a typical micro-cantilever based mechanical microsensor [21]. In addition, the molecular-adsorption-induced stress on a bimaterial microcantilever is used to characterize vapor concentration due to differential stress, resulting in readily measurable curvatures of the cantilever structure [22].With the introduction of measuring and scanning techniques, several deflection detection methods have been proposed to measure microcantilever deflection. The most common and commercially available measuring methods can be divided into optical and electric methods [23-27]. However, the optical approaches cannot effectively get rid of the disadvantages attributed to the bulky laser unit. The electric-based mechanisms including the piezoresitive scheme[24], the piezoelectric scheme[25-27] and the capacitive scheme[28] have been demonstrated about a decade. There are technological limits in fabricating thin and highly sensitive cantilevers using this electric method.

Recently some microcantilever force sensors based on photonic crystal have been reported in Refs. [25-35]. With the advantages of ultracompact size, high contrast and easy integration on photonic crystal, Lee et al. [31] designed and analyzed a novel highly sensitive nano-sacled force sensor. The sensor consists of a micro-cantilever embedded and a nanocavity resonator. They found that the resonant wavelength is a function of the force loading on the surface of the microcantilever. Kramper et al. [30] studies the properties of photonic crystal nanocavity resonators experimentally. They used the scanning near-field optical microscopy (SNOM) to visualize the optical intensity topography of nanocavity photonic crystal resonators. And all the nano-force sensors are in a micro-scaled array.

As can be seen, a lot of work has been done on the field of nano-scaled force sensors, and these force sensors shows good linear characteristics, however, all of the nano-scaled force sensors can only be used to measure one dimensional force. No investigation is found in the literature designing a force sensor which can be used to measure two dimensional forces simultaneously based on the photonic crystal. This two dimensional nano-scaled force sensor can effectively get rid of the disadvantages of attributed to bulky detection unit and provide a feasible way of detecting two dimensional force in micro-scaled array. This force sensor can be widely used in a variety of fields, such as mechanobiology, material science, microrobotics and life science. In this paper, the component force in X and Y direction can be measured by the wavelength and wavelength shift.

Design and modelling of the force sensor

Lee et al.[27] first use a silicon photonic crystal cantilever embedded with a nanocavity resonator as one-dimensional optical nanomechanical sensor. This mechanical sensor is a silicon cantilever comprising a two-dimensional photonic crystal nanocavity resonator arranged in a U-shaped silicon photonica crystal waveguide. In the graph of strain versus resonant wavelength shift, a rather linear relationship is observed for various data derived from different cantilevers. Both the resonant wavelength and the resonant wavelength shift of cantilevers under deformation or force loads are mainly a function of defect length change. The resonant wavelength shift is dominated by the change in defect length of the nanocavity along the longitudinal direction. And all these mechanical parameters are mainly dependent on the defect length of the PC nanocavity resonator.

Fig. 1 (a) presents the side view drawing of a two dimensional nano-scaled force sensor based on silicon photonic crystal. The force sensor is consisting of a silicon substrate frame and two micro-cantilevers embedded with a photonic crystal nanocavity resonator. As we all known, the deflection of the micro-cantilever is sensitive to the changes of the forces loading on it [29].

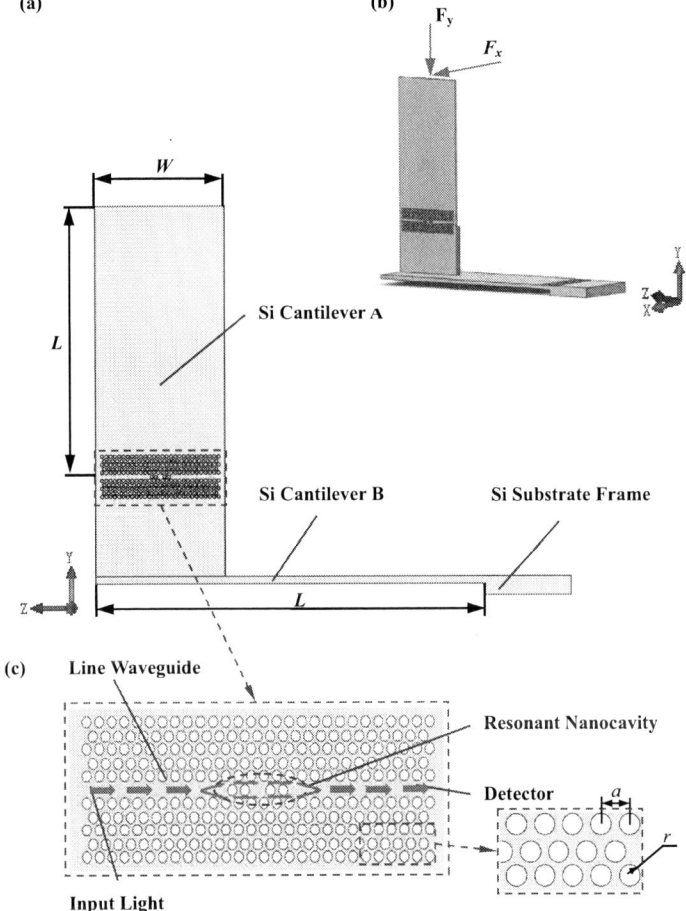

Figure 1 (a) Side view drawing of two dimensional nano-scaled force sensor based on silicon photonic crystal, (b) Tilted top view drawing of two dimensional nano-scaled force sensor based on silicon photonic crystal (c) Schematic drawing of the a nanocavity photonic crystal waveguide resonator on a line waveguide.

The induced micro-cantilever deflection is characterized in terms of the resonant wavelength shift due to deformation of photonic crystal resonator. These photonic crystal cantilevers are varied in terms of length (L) and the width (W), i.e., L/W=30/10, 30/15 and 25/15. The two dimensional solid model is built by a commercial finite element method (FEM) software, i.e., ANSYS 13.0. Fig.1 (b) presents tilted top view drawing of two dimensional nano-scaled force sensor based on silicon photonic crystal. The applied force acts on the top of the nano-scaled force sensor, and the Si substrate frame is fixed. The component force in X and Y directions can be measured by the Si micro-cantilever B and A, respectively. Fig.1 (c) presents schematic drawing of a nanocavity photonic crystal waveguide resonator on a line waveguide. The photonic crystal structure contains a hexagonal array in the silicon micro-cantilever with a lattice constant of a=500nm, and the radius of all holes is r=180nm. The line waveguide is formed by removing a row of airholes. And the nanocavity resonator comprise a one dimensional periodic airholes and a local defect, which fabricated by removing an airhole.

In this design, two pairs of airholes located along a nanocavy photonic resonator. The two of the high-Q resonances centered at wavelengths are 3621 and 3843nm, respectively. It is reported that two of the resonant wavelength were measured as 640 and 190 [31]. Karmper et al. applied scanning near-field optical microscopy to visualize the optical intensity topography around these four airholes of this nanocavity pc resonator. The peak intensity of resonance shown at 3840 nm has been observed [35]. It allows densely arranged photonic waveguides and photonic crystal cantilevers, leading to a feasible sensor array testing setup. Thus, we can apply such a photpmoc array as a novel nanomechanical sensing platform.

Fig 2 (a) presents the strain distribution on nano-scaled force sensor of component force F_x=1μN. In the initial design, all of the thickness of three microcantilevers is 0.2μmThe ratio of strain on nanocavity A and nanocavity B is σ_A/σ_B=2.04×10^5. The strain on nanocavity A is much larger than that of nanocavity B. Hence, the effect of component force in X direction on nanocavity B can be ignored. Fig 2 (b) presents the strain distribution on nano-scaled force sensor of component force F_x=1μN. The ratio of strain on nanocavity A and nanocavity B is σ_A/σ_B=6.59×10^4. The strain on nanocavity A is much larger than that of nanocavity B.

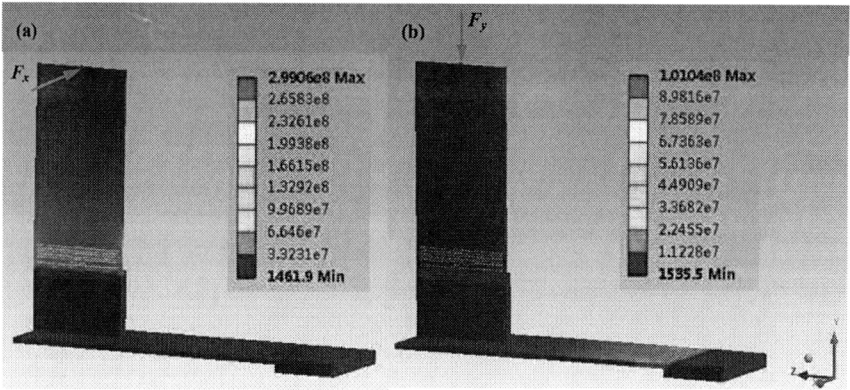

Figure 2 (a) strain distribution on nano-scaled force sensor of component force Fx=1μN, (b) strain distribution on nano-scaled force sensor of component force FY=1μN

Fig. 3 (a) presents the displacement at cantilever end versus different the component force in X direction for a 30 μm long and 15μm wide microcantilever. We observe that the displacement of cantilever end increases with an increase of the applied force in X direction on microcantilever A. However, the displacement tends to 0 μm as the force increases. Hence, only the microcantilever A is sensitive to the component force in X direction. Fig. 3 (b) presents that the displacement of cantilever end versus different the component force in X direction for a 30 μm long and 15μm wide microcantilever. We observe that the displacement of cantilever end increases with an increase of the applied force in Y direction on microcantilever B. However, the displacement tends to 0 μm as the force increases. Hence, only the microcantilever B is sensitive to the component force in X direction. Therefore, the effect of component force in X direction on nanocavity B can be ignored. The component force in X and Y directions can be measured by the Si micro-cantilever B and A, respectively.

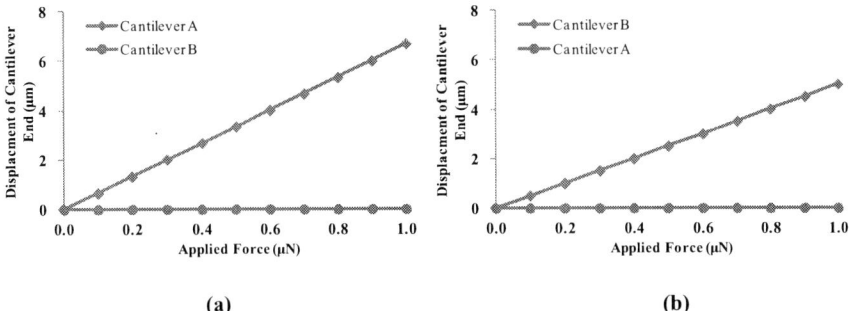

(a)   (b)

Figure. 3 (a) displacement at cantilever end versus different the component force in X direction, (b) displacement at cantilever end versus different the component force in Y direction

Fig 4 (a) presents the fabrication of nanocavity. The nanocavities were fabricated on silicon-on insulator (SOI) wafers using electron beam lithography and reativeion etching. The SOI substrate had a top device layer and a buried oxide layer. The oxide hard mask layer was thermally grown on the Si substrate by a wet oxidation process. Polymethylmethacrylate was then coated on the oxidized substrate as a soft etching mask to write high resolution PhC patterns using an e-beam writing system. After pattern writing and developing, the oxide hard mask layer was dry etched to transfer the patterns into the hard mask. The underlying Si device layer was etched. After completion of the etching process, the PhC devices were finished. Fig. 4 (b) presents a schematic of the measurement setup. Light from the laser source was transmitted through a polarization controller to excite the TE modes and coupled through tapered ridge waveguides into the PhC device using a tapered lensed fiber. The optical power transmitted through the device was collected via a second tapered lensed fiber and measured using an Indium Gallium Arsenide (InGaAs) photodiode detector [36].

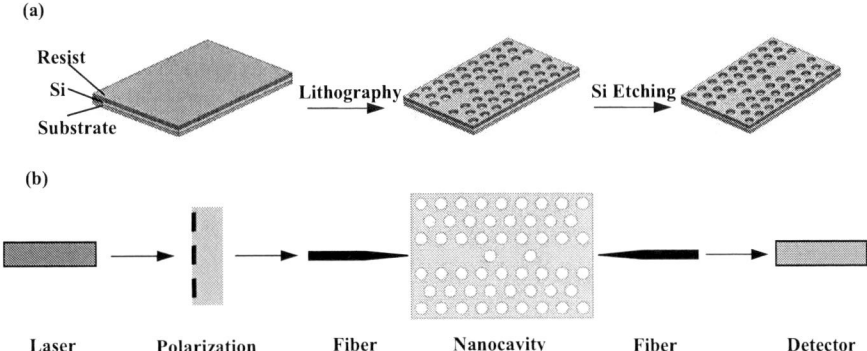

Figure 4 (a) schematic of the fabrication process, (b) schematic of optical detection

Results and discussion

Fig. 5 (a) presents vertical displacement at cantilever end versus different the component force in X direction with different geometries. The force sensor in X direction showed a linear characteristic. Fig. 5 (b) presents vertical displacement at cantilever end versus different the component force in Y direction with different geometries. This force sensor in Y direction showed a linear characteristic.

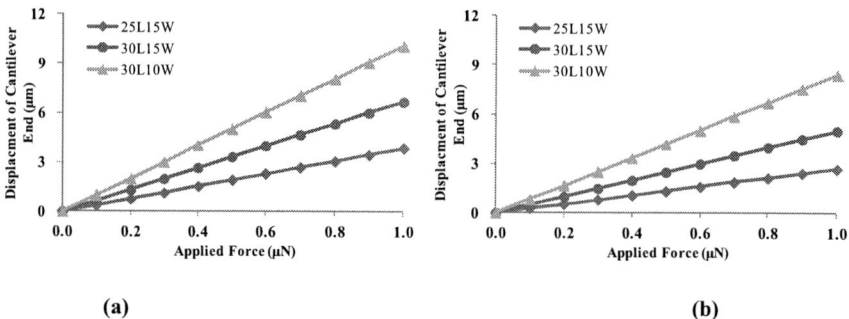

(a)   (b)

Figure 5 (a) displacement at cantilever end versus different the component force in X direction with the different geometries, (b) displacement at cantilever end versus different the component force in Y direction with the different geometries.

Fig. 6 (a) presents resonant wavelength versus different the component force in X direction with the different geometry. The resonant wavelength shifts is measured as a function of the component force in X direction. The force resolutions are 1.500nm/µN, 1.737nm/µN and 2.852 nm/µN for L/W=25/15, 30/15 and 30/10, respectively. 5 (b) presents resonant wavelength versus different the component force in Y direction with the different geometric size. The resonant wavelength is measured as a function of the component force in Y direction. The force resolutions are 0.109nm/µN, 0.126nm/µN and 0.206 nm/µN for L/W=25/15, 30/15 and 30/10, respectively.

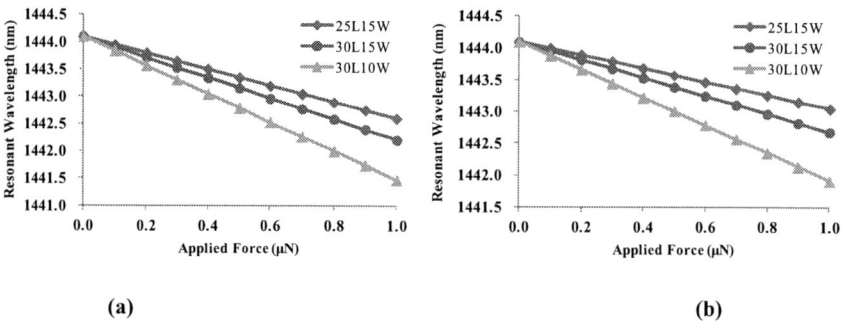

(a)   (b)

Figure 6 (a) resonant wavelength versus different the component force in X direction with the different geometries, (b) resonant wavelength versus different the component force in Y direction with the different geometries

From fig. 6 (a), the relationship between resonant wavelength, Y and applied force X in X direction for L/W=25/15, 30/15 and 30/10 can be shown as:

$$Y=-K_1X+1444.097 \qquad [1]$$

Where K_1 are 1.500, 1.891 and 2.620 for L/W=30/10, 30/15 and 25/15, respectively.

From fig. 6 (b), the relationship between resonant wavelength, Y and applied force X in Y direction for L/W=25/15, 30/15 and 30/10 can be shown as:

$$Y=-K_2X+1444.097 \qquad [2]$$

Where K_2 are 1.050, 1.418 and 2.184 for L/W=30/10, 30/15 and 25/15, respectively.

Fig. 7 (a) presents resonant wavelength shift versus different the component force in X direction with the different geometric size. The wavelength shift, which is mainly attributed to changes in the lattice constant and the radius of hole, can be defined as the difference between two resonant wavelengths. From zero point, the wavelength shift increases with an increase of the applied force. The resonant wavelength shift is measured as a function of the component force in X direction. The force resolutions are 0.109nm/μN, 0.126nm/μN and 0.206 nm/μN for L/W=25/15, 30/15 and 30/10, respectively. Fig. 7 (b) presents resonant wavelength shift versus different the component force in Y direction with the different geometriy. The resonant wavelength shift is measured as a function of the component force in Y direction. The force resolutions are 1.050nm/μN, 1.303nm/μN and 2.377nm/μN for L/W=25/15, 30/15 and 30/10, respectively.

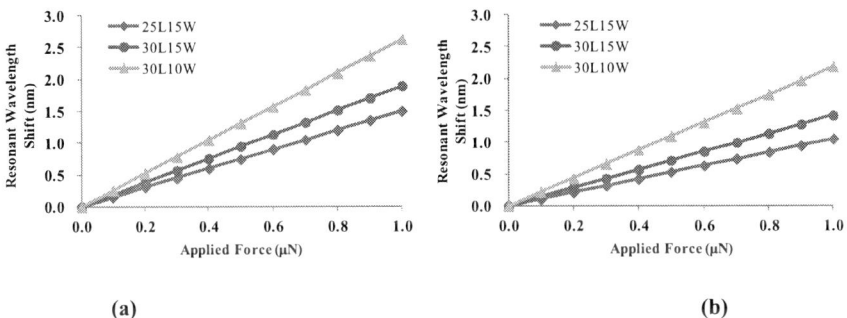

(a) (b)

Figure 7 (a) resonant wavelength shift versus different the component force in X direction with the different geometries, (b) resonant wavelength shift versus different the component force in Y direction with the different geometries

From fig. 7 (a), the relationship between resonant wavelength shift, Y and applied force X in X direction for L/W=25/15, 30/15 and 30/10 can be shown as:

$$Y= K_3X \qquad [3]$$

Where K_3 are 1.500, 1.891 and 2.620 for L/W=25/15, 30/15 and 30/10, respectively.

From fig. 7 (b), the relationship between resonant wavelength shift, Y and applied force X in Y direction for L/W=25/15, 30/15 and 30/10 can be shown as:

$$Y=K_4X \qquad [4]$$

Where K_2 are 1.050, 1.418 and 2.184 for L/W=25/15, 30/15 and 30/10, respectively.

Conclusion

A novel two dimensional nano-scaled force sensor based on silicon photonic crystal which can be used to measure the component force in X and Y directions, respectively was investigated. This novel two dimensional nano-scaled force sensor exhibits an outstanding linear behavior. The resonant wavelength or the resonant wavelength shift is measured as a function of the applied force. The resonant wavelength shift is dominated by the change in defect length of the nanocavity along the longitudinal direction. As we found, the range of the force sensor in each component force in X and Y directions are 0-1µN.And the resolutions of each component force in X and Y directions are 1.891 nm/µN and 1.418 nm/µN, respectively, for a 30 µm long and 15µm wide microcantilever. Results point out that the novel photonic crystal sensor shows promising linear characteristics as an optical nanomechanical sensor.

Acknowledgments

This research is supported by the National Natural Science Foundation of China (51105099), China Postdoctoral Science Foundation (49) and the Postdoctoral Science Foundation of the Heilongjiang Province.

References

[1] E. Yablonovitch, *Phys. Rev. Lett.* 58, 2059-2062 (1987)

[2] S.G. Johnson, S. Fan, P.R. Villeneuve, J. D. Joannopoulos, and L.A. Kolodziejski, *Phys. Rev.* B 60, 5751-5780 (1999)

[3] Chew, Xiongyeu, et al. *Optics express.* 18.21 (2010): 22232-22244.

[4] Nielson, Gregory N., et al. Photonics Technology Letters, IEEE 17.6 (2005): 1190-1192.

[5] Pask, H. M., et al. "Selected Topics in Quantum electronics."

[6] Hsu, Shu-Ting, et al. *MOEMS-MEMS 2008 Micro and Nanofabrication. International Society for Optics and Photonics,* 2008.

[7] Levy O, Steinberg B Z, Nathan M, et al. *Applied Physics Letters,* 2005, 86(10): 104102-104102-3.

[8] Z. Xu, L. Cao, C. Gu, Q. He, and G. Jin. *Opt. Exp,* 14, 298-305 (2006)

[9] Z. Xu, L. Cao, P. Su, Q. He, G. Jin, and G. Gu, *IEEE J Quantum Elect,* 43, 182-187 (2007)

[10] T. Stomeo, M. Grande, A. Qualtieri, A. Passaseo, A. Salhi, M. De Vittorio, D. Biallo, A. D'orazio, M. Sario, V. Marrocco, V. Petruzzelli, F. Prudenzano, *Microelectron Eng,* 84, 1450-1453 (2007)

[11] R. Raiteria, M. Grattarolaa, H. Buttb, P. Skladalc. *Sensors and Actuators B,* 79, 115-126 (2001)

[12] D. R. Baselt, G. U. Lee, K. M. Hansen, L. A. Chrisey, and R. J. Colton, *Proc IEEE,* 85, 672–680 (1997)

[13] N. V. Lavrik, M. J. Sepaniak, and P. G. Datskos, *Rev. Sci. Instrum,* 75, 2229–2253 (2004)

[14] C. Ziegler, Anal Bioanal Chem, 379, 946–959 (2004)

[15] T. R. Albrecht, S. Akamine, T. E. Carver, and C. F. Quate, *J. Vac. Sci. Technol. A,* 8, 3386–3396 (1990)

[16] J. K. Gimzewski, Ch. Gerber, E. Meyer, and R. R. Schlitter, *Chem. Phys. Lett,* 217, 589–594 (1994)

[17] T. Thundat, G. Y. Chen, R. J. Warmack, D. P. Allison, and E. A. Wachter, *Anal. Chem,* vol. 67, pp. 519–521 (1995)

[18] S. J. O'Shea, M. E. Welland, T. A. Brunt, A. R. Ramadan, and T. Rayment, *J. Vac. Sci. Technol.* B 14, 1383–1385 (1996)

[19] S. Asher, V. Alexeev, A. Goponenko, A. Sharma, I. Lednev, C. Wilcox and D. Finegold, J Am. Chem. 125, 3322-3329 (2003)

[20] K. Lee and S. *Asher, J. Am. Chem,* 122, 9534-9537 (2000)

[21] Raiteri, Roberto, et al. *Sensors and Actuators B: Chemical,* 79.2 (2001): 115-126.

[22] Hu, Zhiyu, T. Thundat, and R. J. Warmack. Journal of *Applied Physics,* 90.1 (2001): 427-431.

[23] Meyer, Gerhard, and Nabil M. Amer. *Applied Physics Letters*, 53.12 (1988): 1045-1047.

[24] Tortonese, M., R. C. Barrett, and C. F. Quate. *Applied Physics Letters*, 62 (1993): 834.

[25] Itoh, T., and T. Suga. *Nanotechnology,* 4.4 (1993): 218.

[26] Lee, Chengkuo, et al. *Review of scientific instruments*, 68.5 (1997): 2091-2100.

[27] Lee, Chengkuo, Toshihiro Itoh, and Tadatomo Suga. Sensors and Actuators *A: Physical* 72.2 (1999): 179-188.

[28] Mai, Trong Thi, et al. Sensors and Actuators A: Physical 165.1 (2011): 16-25.

[29] C. Lee and J. Thillaigovindan, *Appl Optics,* 48, 1797-1803 (2009)

[30] P. Kramper, M. Kafesaki, C. M. Soukoulis, A. Birner, F. Muller, R. Wehrspohn, U. Gosele, J. Mlynek, and V. Sandoghdar, *Opt. Lett.* 29, 174–176 (2004)

[31] C. Lee, C. Chen, J. Li, J. Thillaigovindan and N. Balasubramanian, *J Lightwave Technol* 26, 839-846 (2008)

[32] P. Kramper, A. Birner, M. Agio, C Soukoulis, F. Muller, U. Gosele, J. Mlynek and V. Sandoghdar, *Phys Rev B,* 64, 1-4 (2001)

[33] W. Xiang and C. Lee, IEEE J. Sel. Top Quant. 15, 1323-1326 (2009)

[34] T. Maia, F. Hsiaoa, C. Lee, W. Xiang, C. Chen, W. Choi, *Sensor Actuat.* A 165, 16-25 (2011)

[35] Kramper, Patrick, et al. *Optics letters*, 29.2 (2004): 174-176.

[36]Pal, Sudeshna, et al. *Biosensors and Bioelectronics,* 26.10 (2011): 4024-4031.

74

Author Index

Arslan, M.	33	Zhang, G.	65
		Zhu, Y.	41
Belghiti, D.	55		
Birdsong, T.	33		
Biswal, D.	33		
Braustein, H. E.	1		
Braustein, I. E.	1		
Chouarbi, K.	55		
Cortés, M.	41		
Couty, M.	41		
Dinh, T. H. N.	41		
Dondapati, H.	33		
Garel, O.	41		
Lefeuvre, E.	41		
Li, L.	65		
Li, T.	65		
Li, Y.	65		
Moulin, J.	41, 55		
Peng, T.	41		
Pradhan, A. K.	33		
Samantaray, C.	33		
Santiago, K.	33		
Smyrl, W. H.	19		
Song, W.	65		
Souadda, M.	41		
Woytasik, M.	41, 55		